Revision Notes on Construction Science

P. J. Innocent

B.Sc., M.Illum.E.S., M. I. Env. S.

Department of Construction and Environmental Health,
Bristol Polytechnic

NEWNES-BUTTERWORTHS

LONDON BOSTON

Sydney Wellington Durban Toronto

THE BUTTERWORTH GROUP

UNITED KINGDOM Butterworth & Co (Publishers) Ltd.
London: 88 Kingsway, WC2B 6AB

AUSTRALIA Butterworths Pty Ltd.
Sydney: 586 Pacific Highway, Chatswood, NSW 2067
Also at Melbourne, Brisbane, Adelaide and Perth

CANADA Butterworth & Co (Canada) Ltd.
Toronto: 2265 Midland Avenue, Scarborough, Ontario M1P 4S1

NEW ZEALAND Butterworths of New Zealand Ltd.
Wellington: 26-28 Waring Taylor Street, 1

SOUTH AFRICA Butterworth & Co (South Africa) (Pty) Ltd.
Durban: 152-154 Gale Street

USA Butterworth (Publishers) Inc.
Boston: 19 Cummings Park, Woburn, Mass. 01801

First published 1977

ISBN 0 408 00279 4

Typeset by Reproduction Drawings Ltd., Sutton, Surrey.

Printed in England by J. W. Arrowsmith Ltd., Bristol

Preface

The aim of this book is to provide a concise account of the fundamental principles of physics and materials science related to the technology of the built environment. Such a short volume could not furnish a fully comprehensive text on this subject, and thus it will be necessary for students to complement these notes with additional detail and information gained from lectures and background reading, in addition to consolidation of this knowledge by its application to appropriate tutorial problems. Unfortunately, lack of space precludes the inclusion of numerous worked examples and self-testing questions.

I have deliberately minimised the mathematical content included in the book, so that having mastered the basic concepts one can, with appropriate mathematical extensions, elucidate their more complex interrelationships.

Physics is very much a science of measurement, and thus analysis and evaluation of the results obtained from judiciously chosen laboratory experiments should provide an adequate synthesis of theoretical concepts, and practical and intellectual skills.

S.I. units (Le Système International d'Unités) and symbols have been used in this text, with the current nomenclature of physical quantities and terminology as recommended by the Symbols Committee of the Royal Society (London).

I am grateful to the Illuminating Engineering Society for permission to reproduce figures from their Technical Report No 2; to Bruel and Kjaer, Denmark, for the Equal Loudness Contours and Index graphs, Perceived Noisiness Contours and Noise Rating curves; and to Cuprinol Ltd for extracts from their technical literature on timber preservation.

Additional sources of reference:

British Standards, British Standards Institution
British Standard Codes of Practice, British Standards Institution
The Building Regulations, H.M.S.O.
Building Research Establishment Digests, H.M.S.O.
Department of the Environment Advisory Leaflets, H.M.S.O.
Chartered Institution of Building Services (formerly I.H.V.E.) *Guide*
Illuminating Engineering Society (I.E.S.) *Code*
Specification, Architectural Press

Contents

Physical quantities

Physical quantity = numerical value × unit

The International System of Units (SI)

The International System of Units (SI) has been established by resolutions of the General Conference on Weights and Measures (CGPM).

There are 3 types of units: *base, supplementary* and *derived.*

Physical quantity	*Symbol*	*SI unit*	*Symbol*
SI base units			
length	l	metre	m
mass	m	kilogram	kg
time	t	second	s
electric current	I	ampere	A
thermodynamic temperature	T	kelvin	K
luminous intensity	I	candela	cd
amount of substance	n	mole	mol
SI supplementary units			
plane angle	θ, ϕ	radian	rad
solid angle	Ω, ω	steradian	sr
Examples of SI derived units			
area	A	square metre	m^2
volume	V	cubic metre	m^3
density	ρ	kilogram per cubic metre	$kg\,m^{-3}$
speed, velocity	u, v	metre per second	$m\,s^{-1}$
angular velocity	ω	radian per second	$rad\,s^{-1}$
acceleration	a	metre per second squared	$m\,s^{-2}$
surface tension	γ	newton per metre	$N\,m^{-1}$
heat capacity	C	joule per kelvin	$J\,K^{-1}$
thermal conductivity	k	watt per metre kelvin	$W\,m^{-1}\,K^{-1}$

Note: 1. The use of the word 'specific' before the name of a physical quantity means 'divided by mass' (e.g. specific heat capacity).
2. Terms ending in '-ivity' refer to a unit thickness of a material and the symbols used are *lower case* letters (e.g. thermal conductivity, k).
3. A solidus (/) and parentheses may alternatively be used to denote SI derived units, e.g. W/(m K).

2 QUANTITIES, UNITS & SYMBOLS (2)

SI derived units with special names

		SI unit			
Physical quantity	*Symbol*	*Name*	*Symbol*	*Definition*	*Equivalent form(s)*
energy	E	joule	J	$m^2\,kg\,s^{-2}$	N m
force	F	newton	N	$m\,kg\,s^{-2}$	$J\,m^{-1}$
pressure	p	pascal	Pa	$m^{-1}\,kg\,s^{-2}$	$N\,m^{-2}$, $J\,m^{-3}$
power	P	watt	W	$m^2\,kg\,s^{-3}$	$J\,s^{-1}$
electric charge	Q	coulomb	C	s A	A s
electric potential difference	V	volt	V	$m^2\,kg\,s^{-3}\,A^{-1}$	$J\,A^{-1}\,s^{-1}$
electric resistance	R	ohm	Ω	$m^2\,kg\,s^{-3}\,A^{-2}$	$V\,A^{-1}$
electric conductance	G	siemens	S	$m^{-2}\,kg^{-1}\,s^3\,A^2$	Ω^{-1}, $A\,V^{-1}$
electric capacitance	C	farad	F	$m^{-2}\,kg^{-1}\,s^4\,A^2$	$A\,s\,V^{-1}$
magnetic flux	Φ	weber	Wb	$m^2\,kg\,s^{-2}\,A^{-1}$	V s
magnetic flux density	B	tesla	T	$kg\,s^{-2}\,A^{-1}$	$Wb\,m^{-2}$, V s m
inductance	L	henry	H	$m^2\,kg\,s^{-2}\,A^{-2}$	$V\,A^{-1}\,s$
luminous flux	Φ	lumen	lm	cd sr	
illuminance	E	lux	lx	m^{-2} cd sr	
frequency	f, ν	hertz	Hz	s^{-1}	
absorbed dose (radiation)		gray	Gy	$J\,kg^{-1}$	
activity (radioactive source)	A	becquerel	Bq	s^{-1}	

SI prefixes

Note that decimal multiples of the kilogram should be formed by attaching an SI prefix to the gram, not to the kilogram.

Multiple	*Prefix*	*Symbol*
10^{-1}	deci	d
10^{-2}	centi	c
10^{-3}	milli	m
10^{-6}	micro	μ

Multiple	*Prefix*	*Symbol*
10	deca	da
10^2	hecto	h
10^3	kilo	k
10^6	mega	M

Mathematical symbols and formulae

$+$	plus
$-$	minus
$\pm$	plus or minus
$=$	is equal to
$\neq$	is not equal to
$\approx$	is approximately equal to
$<$	is less than
$>$	is greater than
$\leqslant$	is less than or equal to
$\geqslant$	is greater than or equal to
$\ll$	is much less than
$\gg$	is much greater than

a^n	a raised to power n
$\sqrt{a}$, $a^{\frac{1}{2}}$	square root of a
$\lg x$, $\log_{10} x$	common logarithm of x
$\ln x$, $\log_e x$	natural logarithm of x
Δx	finite increment of x
δx	variation of x
$\mathrm{d}f/\mathrm{d}x$	differential coefficient of $f(x)$ with respect to x

List of derivatives

y	$\mathrm{d}y/\mathrm{d}x$
$\sin x$	$\cos x$
$\cos x$	$-\sin x$
$\tan x$	$\sec^2 x$
$\sec x$	$\sec x \tan x$

Mensuration formulae

Lines: Pythagorean Theorem $a^2 = b^2 + c^2$
circumference of a circle $= 2\pi r$

Plane areas: triangle $= \frac{1}{2}bh$
parallelogram $= bh$

Curved surfaces: cylinder $= 2\pi rh$
sphere $= 4\pi r^2$

Volumes: cylinder $= \pi r^2 h$
sphere $= \frac{4}{3}\pi r^3$

Constants

$\pi \approx 3.141\,593$ $\quad$ $e \approx 2.718\,281$

Matter

3 states (solids, liquids and gases).
Composed of *elements* and *compounds.*

Elements

Basic 'building blocks' of matter (e.g. H, O, Fe, Hg).

Atoms

Smallest part of an element that can take part in a chemical change.
Atoms of different elements have different masses.
Atoms consist of smaller particles, fundamental ones being *protons, neutrons* and *electrons.* Properties:

Particle	*Mass*	*Charge* (coulomb)
proton	1	$+1.6 \times 10^{-19}$
neutron	1	0
electron	1/1836	-1.6×10^{-19}

Atomic mass unit (u) = 1.66×10^{-27} kg

Molecule

Smallest part of a substance that can exist in a free state.
May contain one or more atoms (e.g. H_2).

Compound

Chemical combination of two or more elements to form a molecule of a new substance with unique properties (e.g. H_2O).

Mixture

Substances that exist together but do not combine to form a new substance.

Atomic number

Number of protons in an atom.
Identifies a particular element (e.g. Carbon 6).

Mass number

Number of protons and neutrons in atomic *nucleus.*

Specification of a nuclide

$^{A}_{Z}X_{N}$

where X is the chemical symbol
A is the mass number
N is the number of atoms/molecules
Z is the atomic number (not always shown).
The right superscript position indicates ionic charge (e.g. Na^+, Zn^{2+}, Cl^-, $SO_4{}^{2-}$) or state of excitation (e.g. electronic excited state, He*).

Electron orbitals

Number of *positive* charges must equal number of *negative* charges in a neutral atom.
Discrete energy levels (*shells*) in which the electrons orbit the nucleus.

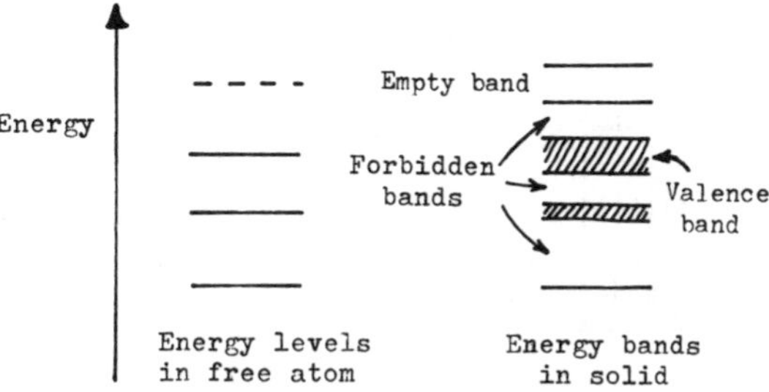

Maximum number of electrons per shell.
Inner (K) shell has lowest energy state and fills up first.
Size of atom approx. diameter of outer electron orbit. Most of the atom is entirely empty space.
Electrons in the outer orbits of some atoms not tightly bound to the nucleus can move from one atom to another (*free electrons*).
Materials with free electrons are electrical *conductors.*

Ionisation

Atoms which gain or lose an electron form *negative* or *positive* ions.

Valency

Number of electrons that an element must lose or gain to form a completed outer shell.

Periodic table

Grouping of elements according to their properties.

4 ATOMIC STRUCTURE (2)

Spectra

Energy (E) is emitted in *quanta* when an electron in an *excited* atom falls to a lower energy level.

$$E = h\nu$$

where h is Planck's constant
ν is the frequency of emitted radiation.

Radiation from *solids: continuous* spectra.

Radiation from *gases and vapours: line* spectra (from atoms) and *band* spectra (from molecules).

Hydrogen atom energy levels
(1 electronvolt (eV) $\approx 1.602 \times 10^{-19}$ J)

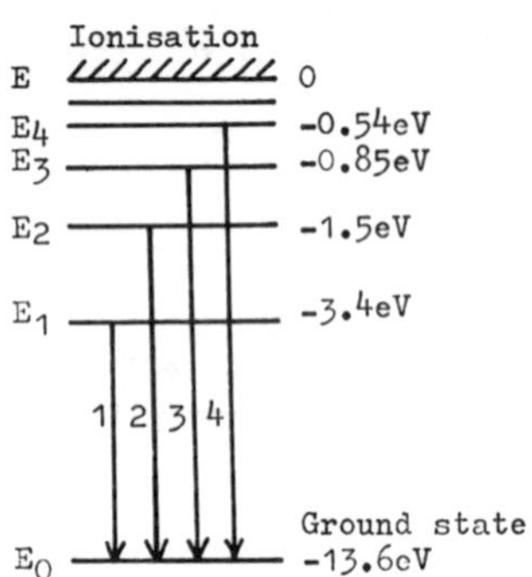

Spectral lines

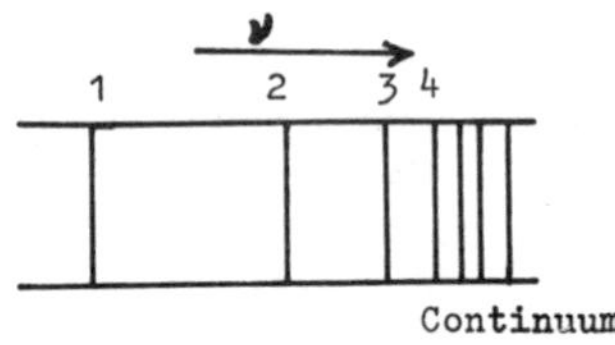

Isotope

An atom whose nucleus has the same number of *protons* as another atom but a different number of *neutrons*.

Both atoms have same chemical properties.

Radioactive isotopes

Unstable atoms that *decay*.

Radioactive decay

Activity, A (becquerel, Bq)
The number of nuclear transformations per unit time of a radioactive source.

$$A = -\frac{dN}{dt}$$

where N is the number of nuclei present
t is time.

Decay constant, λ
The probability per unit time for the decay of an unstable nucleus.

$$\lambda = \frac{A}{N} = -\frac{dN}{dt} \cdot \frac{1}{N}$$

$$N = N_o \exp(-\lambda t) \quad \text{(by integration)}$$

where N_o are the number of nuclei present at time $t = 0$.

Half-life, $T_{\frac{1}{2}}$ (seconds, hours, days, years)
The time taken for half the original number of nuclei to decay ($N = \frac{1}{2}N_o$).

$$T_{\frac{1}{2}} = \frac{\ln 2}{\lambda} = \frac{0.693}{\lambda}$$

where $\ln 2 = \log_e 2$

Forms of radioactivity

Property	*α particle*	*β particle*	*γ radiation*	*Neutron radiation*
Structure	nucleus of helium atom	electron	electromagnetic radiation	neutrons
Mass (a.m.u.)	4	1/1836		1
Charge	2+	1−	nil	nil
Range	3–4 mm in air	2–4 mm in Al	depends on energy	6 m in concrete

Nuclear reactions

initial nuclide $\left(\text{incoming particle(s), or quanta} \quad \text{outgoing particle(s) or quanta}\right)$ final nuclide

e.g. $^{14}N(\alpha, p)^{17}O$

Solid: a state of matter possessing both definite shape and definite volume.

Crystalline solids

Rigid crystalline structure with a definite melting point

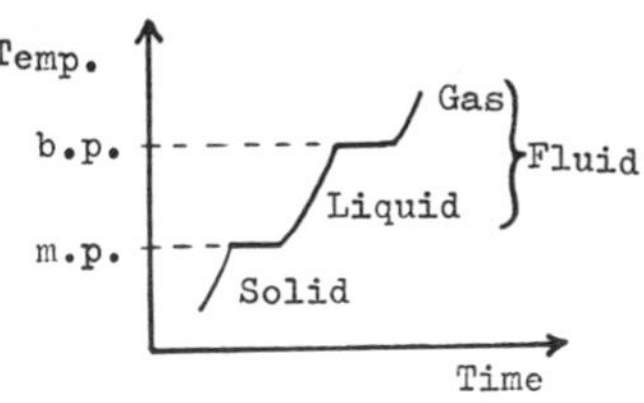

Three types: electrovalent, covalent and metallic crystals.

(a) *Electrovalent crystals*: formed by ions linked by electrovalent *bonds* in a *lattice* structure. These bonds dissolve readily in water producing a dispersion of ions.
(b) *Covalent crystals: either* 'building block'-type structure (e.g. most organic compounds), weak van der Waals forces; *or* 'giant molecule'-type structure (e.g. diamond), strong covalent forces.
(c) *Metallic crystals*: ionised metal atoms positively charged in a lattice but which contains *free* electrons.

Amorphous solids

Non-crystalline structure with no definite melting point (supercooled liquid).

MECHANICS

Mass, m (kilogram, kg): scalar quantity.
Measure of the inertia of a body.
1 tonne (t) = 10^3 kg = 1 Mg.

Density, ρ (kg m^{-3}): scalar quantity.
Mass per unit volume of a substance, i.e. $\rho = m/V$.

Relative density, d: scalar quantity.
Ratio of density of substance (ρ_2) to density of water (ρ_1) at 4°C (= 1000 kg m^{-3}), i.e. $d = \rho_2/\rho_1$.

Velocity, v (m s^{-1}): vector quantity.
Distance moved per unit time in a given direction, i.e. $v = s/t$.

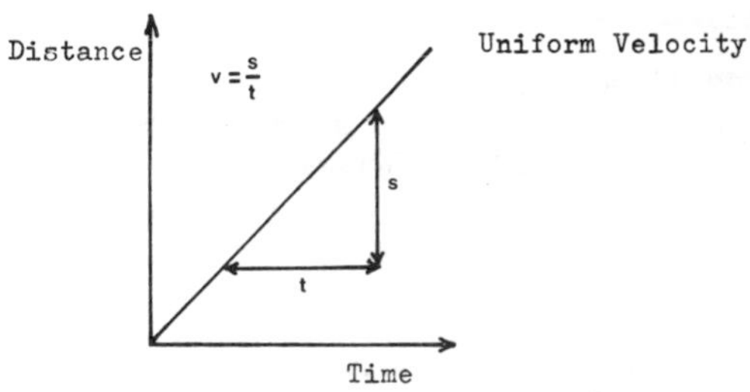

Momentum, p (kg m s^{-1}): vector quantity.
Mass × velocity, i.e. $p = mv$.

Law of conservation of linear momentum

When objects collide, the sum of the momenta before impact equals the sum of the momenta after impact provided no external forces act on the system. (In calculations one direction only is positive.)

Acceleration, a (m s^{-2}): vector quantity.
Rate of change of velocity, i.e. $a = dv/dt$.

Force, F (newton, N): vector quantity.
Force exerted on *unrestricted* object: its *velocity* changes.
Force exerted on *restricted* object: its *shape* changes.

Resolution of a force into two components:

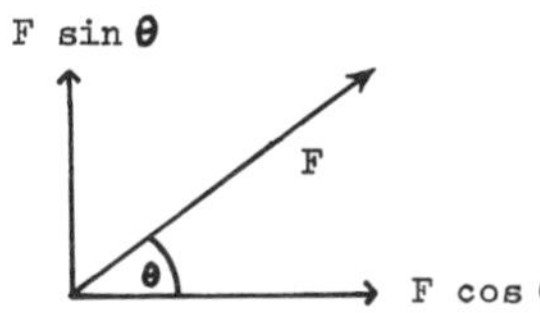

Newton's Laws of Motion

1. Every body continues in its state of rest or uniform motion in a straight line unless acted on by a force.
2. The *rate* of change of momentum (mv) of a body is proportional to the impressed force (F) and takes place in the direction of the force.

$$F \propto \frac{d(mv)}{dt} \approx \frac{mv - mu}{t} = \frac{m(v-u)}{t} = ma$$

The *newton* (N) is the force which produces an acceleration of 1 m s^{-2} in a mass of 1 kg.

3. When a force acts on a body, a force equal in magnitude and opposite in direction (*reaction*) acts on another body.

6 SOLIDS (2)

STATICS

Weight, W (newton, N)
The force exerted by a body when it is supported at the earth's surface. The gravitational force would cause the mass to accelerate at a constant rate if it were free-falling.

$$W = mg$$

where m is the mass (kg)
g is the *acceleration due to gravity* ($\mathrm{m\,s^{-2}}$)
Acceleration due to gravity varies around the world (approx. 9.81 $\mathrm{m\,s^{-2}}$).

Pressure, p (pascal, Pa)
The *normal* force (newton) per unit area ($\mathrm{m^2}$).

$$p = F/A$$

where A is the area.

Tensile forces

Tensile stress = $\sigma = F/A$
Tensile strain = $\epsilon = \Delta l/l_0$
Modulus of elasticity (Young's modulus) = $E = \sigma/\epsilon$

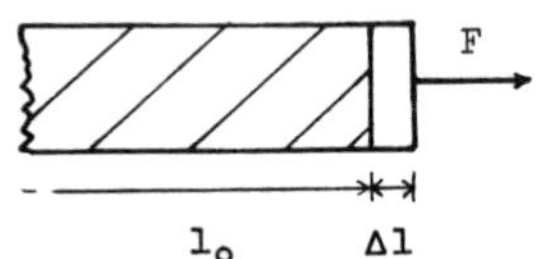

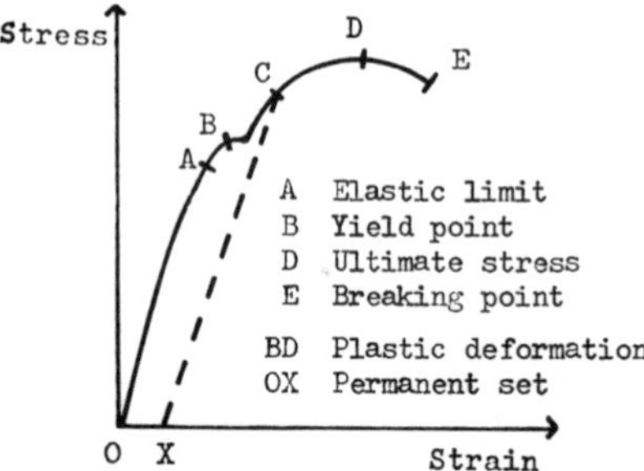

Stress/strain curves for elastic materials

Shear forces

Shear stress = $\tau = F/A$
Shear strain = $\gamma = \Delta x/d$
Modulus of rigidity (shear modulus) = $G = \tau/\gamma$

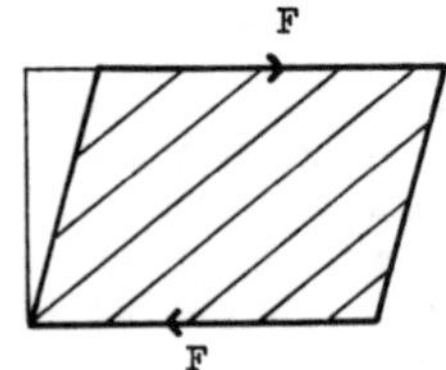

Compressive forces

Bulk stress = $p = F/A$
Bulk strain = $\theta = -\Delta V/V$
Bulk modulus = $K = p/\theta = -p/(\Delta V/V)$

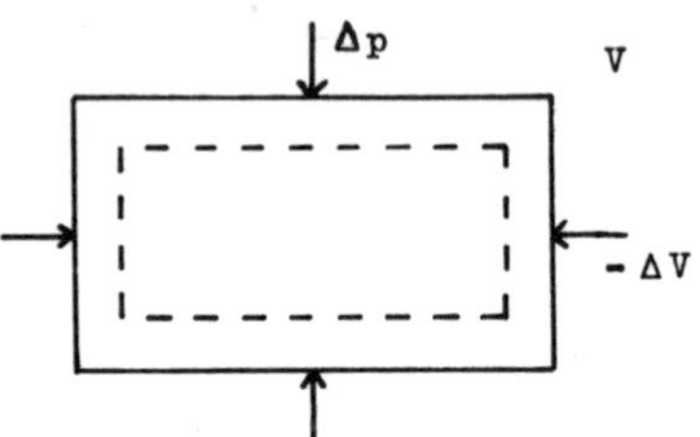

Moment of a force, M (N m)
Moment of force F about point O = Fd

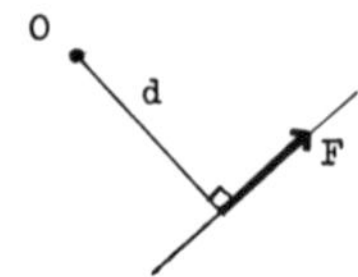

Principle of moments

If a body is in *equilibrium* under the action of *co-planar* forces, then the sum of the *clockwise* moments about a point is equal to the sum of the *anti-clockwise* moments about that point.

DYNAMICS

Mechanical energy

Potential energy, E_p (joule, J)
Energy by virtue of position:

$$E_\mathrm{p} = mgh$$

Kinetic energy, E_k (J)
Energy by virtue of motion:

$$E_\mathrm{k} = \tfrac{1}{2}mv^2$$

Translational, rotational and vibrational.

Principle of Conservation of Energy
The total energy in a given system is constant, although energy may change from one form to another.

Power, P (watt, W)
Rate of expenditure of energy (1 watt = 1 J/s).

Motion

Linear motion (uniform acceleration, a)

Acceleration = gradient of line = a
Total distance travelled = area under graph = s

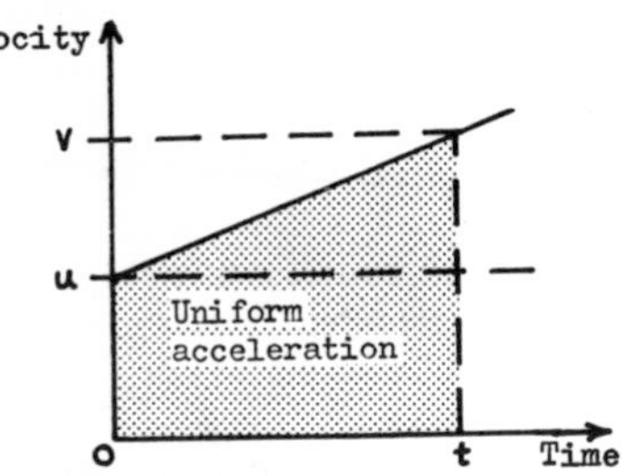

$$v = u + at$$

$$s = ut + \tfrac{1}{2}at^2$$
$$v^2 = u^2 + 2as$$

If an object falls under gravity, $a = +g$
If an object is thrown vertically upwards, $a = -g$

Circular motion

Angular velocity, ω (rad s^{-1})
ω = angular change per second = $d\theta/dt$
Object making f revolutions per second:
$\omega = 2\pi f$

Angular acceleration, α (rad s^{-2})
α = time rate of increase of angular velocity
$= d\omega/dt$

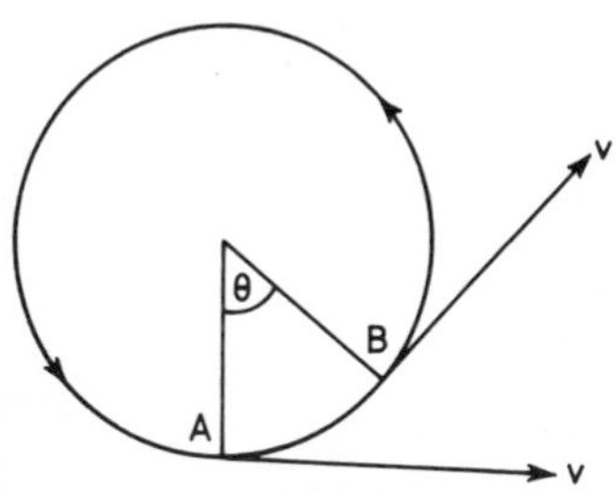

Velocity in circle, v (m s^{-1})
$v = r\omega$

Acceleration in circle, a (m s^{-2})

$$a = \frac{\text{velocity change}}{\text{time}} = \frac{v\theta}{t} = v\omega = \frac{v^2}{r}$$

Acceleration is *towards* centre of circle.

Centripetal force, F (N)
F = force towards centre = $mv^2/r = mr\omega^2$

Linear motion (fixed direction)	*Rotational motion (fixed axis)*
Displacement, s (m)	Angular displacement, θ (rad)
Speed, $u = ds/dt$ (m s^{-1})	Angular speed, $\omega = d\theta/dt$ (rad s^{-1})
Acceleration, $a = du/dt$ (m s^{-2})	Angular acceleration, $\alpha = d\omega/dt$ (rad s^{-2})
$v = u + at$ $s = ut + \frac{1}{2}at^2$ $v^2 = u^2 + 2as$	$\omega_t = \omega_o + \alpha t$ $\theta = \omega_o t + \frac{1}{2}\alpha t^2$ ${\omega_t}^2 = {\omega_o}^2 + 2\alpha\theta$

Simple harmonic motion (S.H.M.)

Angular velocity = ω
Period = $T = 2\pi/\omega$ (1 complete revolution)
Displacement from fixed point = $y = r \sin \omega t$
Velocity at any instant = $v = dy/dt = r\omega \cos \omega t$
$= \pm\omega^2 \sqrt{(r^2 - y^2)}$
Acceleration towards centre = $a = d^2y/dt^2$
$= -r\omega^2 \sin \omega t = -\omega^2 y$
i.e. acceleration is proportional to displacement.

Maximum and minimum values during S.H.M.

Concept	*Position 2* ($\omega t = \pi/2$)	*Position 3* ($\omega t = 0$)
y	max ($= r$)	min ($= 0$)
v	min ($= 0$)	max ($= r\omega$)
a	max ($= -\omega^2 r$)	min ($= 0$)

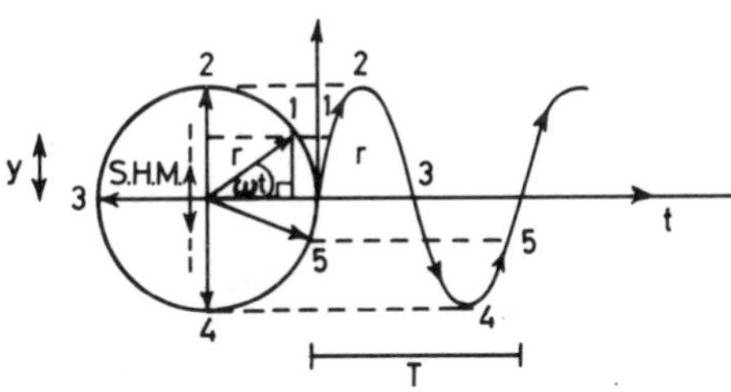

Motion of rigid bodies

Moment of inertia = $I = \Sigma mr^2 = Mk^2$
where k is the *radius of gyration.*

Friction

Coefficient of static friction, μ:
Limiting static frictional force ÷ normal reaction

Coefficient of dynamic friction, f:
Frictional force when one surface is moving over the other ÷ normal reaction.

8 LIQUIDS (1)

HYDROSTATICS

Hydrostatics is the study of liquids at rest.

1. Liquids take any shape.
2. Liquids are nearly incompressible.
3. Pressure at any point in a liquid at rest is exerted equally in all directions.
4. Pressure at any point in a liquid depends only on the vertical height of the head of liquid and not the shape.

Force at point in liquid, F (N)

$$F = mg = Ah\rho g$$

Pressure, p (Pa)
The force exerted per unit area on an infinitesimal plane at a point in a fluid.

$$p = F/A = h\rho g$$

where ρ = density of liquid (kg m^{-3})
g = acceleration due to gravity (9.81 m s^{-2})
h = static head (m)

Surface tension

Molecular forces

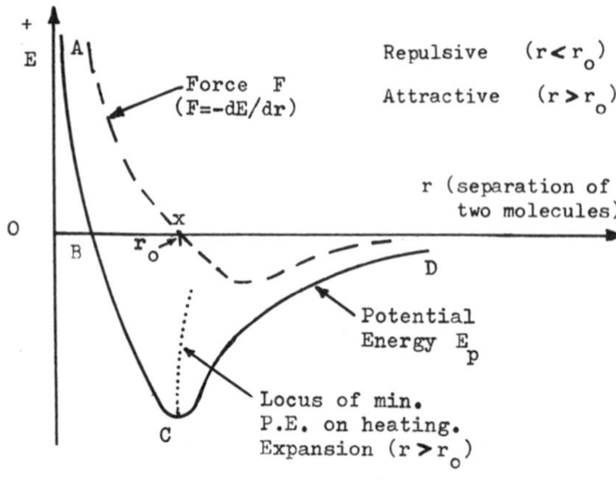

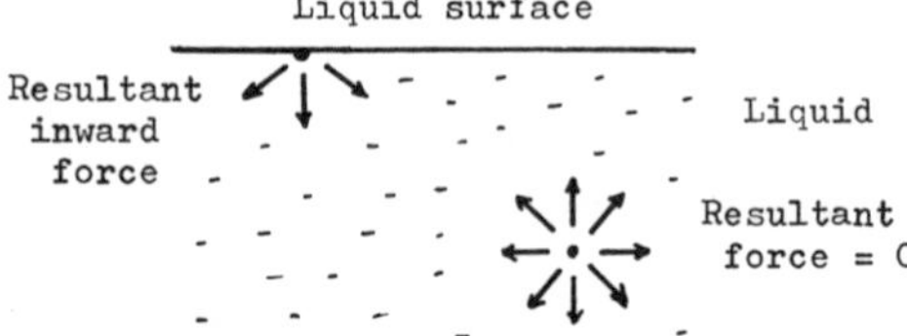

Coefficient of surface tension, γ (N m^{-1})
The force per unit length in the surface acting on one side of, and perpendicular to, a line drawn in the surface.
γ decreases with (a) rise in temperature, (b) impurities in liquid.

Angle of contact, θ

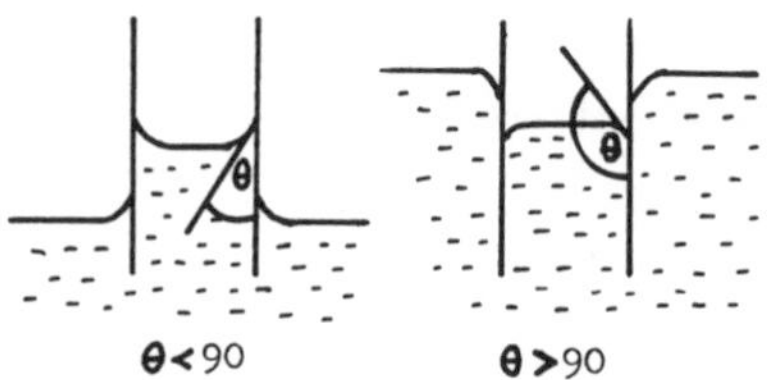

Capillarity

Cause: *adhesion* of liquid/solid (wetting) > *cohesion* of liquid molecules.
Liquids with *acute* angles of contact rise in tubes (e.g. water).
Liquids with *obtuse* angles of contact are depressed in tubes (e.g. mercury).

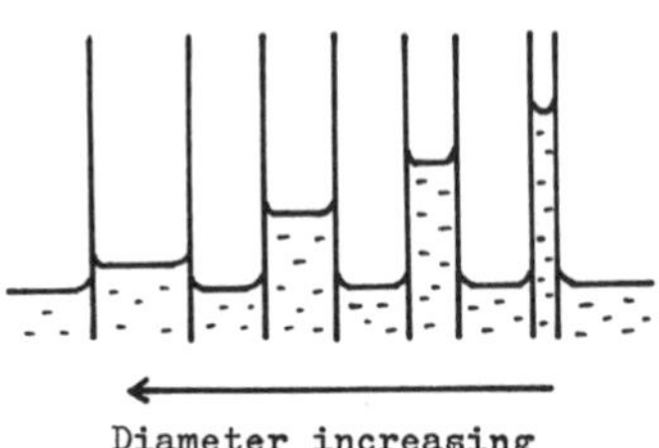

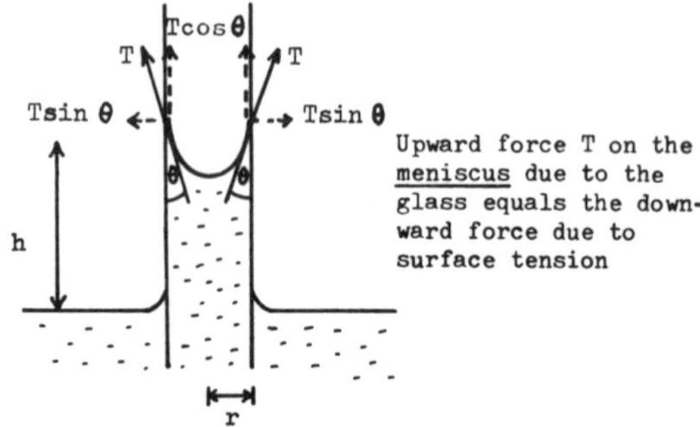

For equilibrium conditions:
(a) resolved *horizontal* forces cancel
(b) *upward* forces = *downward* forces
= volume × density × g

$$2\pi r\gamma\cos\theta = \pi r^2 h\rho g$$

Capillarity in buildings

1. Capillary movement of moisture through porous materials. Small e.m.f. may be generated – electro-osmosis.
2. Passage of water between overlapping roof tiles if no provision for elimination of capillarity paths.
3. Capillary joints.

HYDRODYNAMICS

Hydrodynamics is the study of liquids in motion.

1. Each element of a fluid experiences stresses at its surface due to the effect of other elements of the fluid.
2. Resolved stress component *normal* to direction of fluid motion: *pressure* (occurs in fluids at rest or moving).
3. Resolved stress component *tangential* to direction of fluid motion: *shear stress* (only occurs in moving fluids).
4. Shear stresses are due to *viscosity* of fluid.

Newton's Law. The shearing stress at any point is directly proportional to the velocity gradient at the point perpendicular to the plane considered (applies to streamline flow *only*).

$$\text{Shearing stress} = \frac{\text{viscous drag force}}{\text{area}} = \frac{F}{A}$$

$$\text{Velocity gradient} = \frac{\text{velocity change}}{\text{distance}} = \frac{du}{dy}$$

$$\boxed{\frac{F}{A} = \eta \frac{du}{dy}}$$

where η = *coefficient of dynamic viscosity* (Pa s)

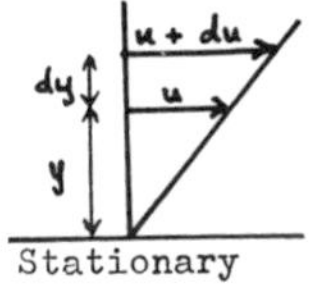

flow of liquid over a flat surface

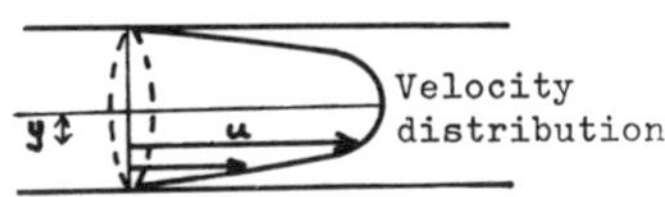

flow of liquid through a pipe

Reynolds number, *Re* (approx. 1000 for narrow tubes):

$$\boxed{\text{Critical velocity} = v_c = \frac{Re\eta}{\rho a}}$$

where ρ = density of liquid (kg m^{-3})
a = lateral dimensions; for pipes a = radius (m)
η/ρ = *kinematic viscosity* (m^2 s^{-1}).

Effects of viscosity in fluid flow through circular pipes:
(a) below critical velocity: *laminar* flow
(b) above critical velocity: *turbulent* flow

Flow in pipes

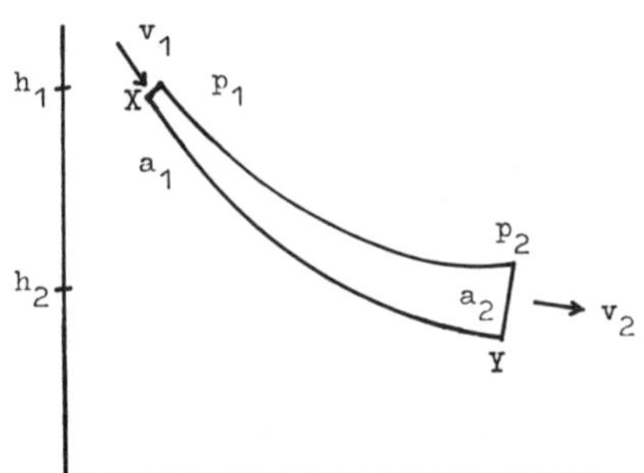

Streamlines of flow of a liquid in a gravitational field

where h_1, h_2 = static pressure heads above reference datum
p_1, p_2 = pressures
v_1, v_2 = velocities of liquid.

Bernoulli's Principle. The sum of the kinetic energy per unit volume, potential energy per unit volume and pressure is constant at any point in an incompressible, non-viscous fluid moving with a uniform or streamline motion.

$$\boxed{\tfrac{1}{2}\rho v_1^2 + \rho g h_1 + p_1 = \tfrac{1}{2}\rho v_2^2 + \rho g h_2 + p_2}$$

where ρ = density of liquid (kg m^{-3}).
Alternatively, dividing by ρg,

$$\boxed{\frac{v_1^2}{2g} + h_1 + \frac{p_1}{\rho g} = \frac{v_2^2}{2g} + h_2 + \frac{p_2}{\rho g}}$$

In practice, some energy is degraded into heat.
Static head = $p/\rho g$
Velocity head = $v^2/2g$

Water hammer. Rapid closure of a valve brings water to rest and, according to Bernoulli's Principle, energy cannot be destroyed but is converted into a shock wave and released as strain energy.

Hydraulic gradient
Rate of discharge from pipe depends on:
(a) static head
(b) frictional resistance to flow
(c) size and shape of orifice
Hydraulic gradient increases as rate of draw-off increases.

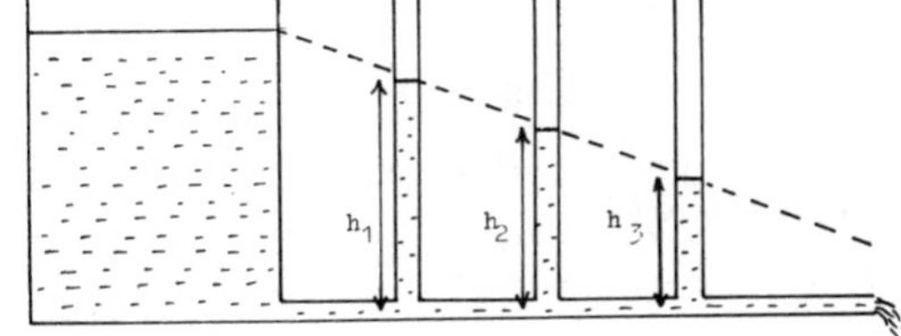

10 LIQUIDS (3)

Flow in open channels

Chézy formula: $v = c\sqrt{(mi)}$

where v = velocity (m/s)
i = hydraulic gradient
c = Chézy constant ($m^{0.5} s^{-1}$)
m = hydraulic mean depth (m), i.e. cross-sectional area of flow ÷ wetted perimeter.

Assumes:
(a) long channel of constant cross-section
(b) roughness of channel walls is constant
(c) gradient of channel bed is constant

Discharge through small orifices

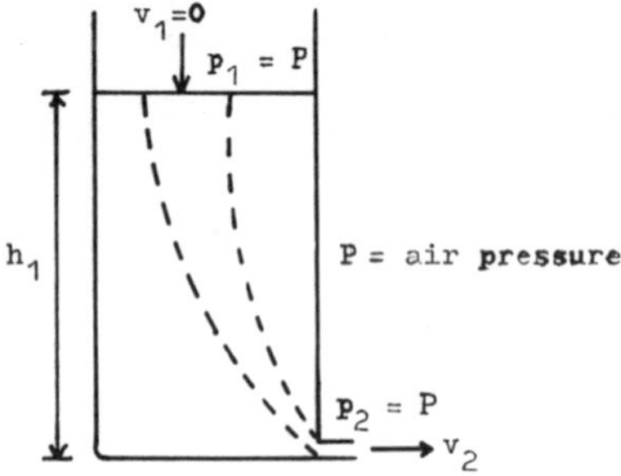

Torricelli's Theorem. The velocity of liquid emerging at a depth h from a hole in a wide vessel open to the air = $\sqrt{(2gh)}$, derived from Bernoulli Principle.

Actual velocity = $C_v \sqrt{(2gh)}$

where C_v = *coefficient of velocity.*

Coefficient of contraction, C_c

$C_c = A_1/A_3$

where A_1 = sectional area of flow at vena contracta
A_3 = sectional area of orifice.

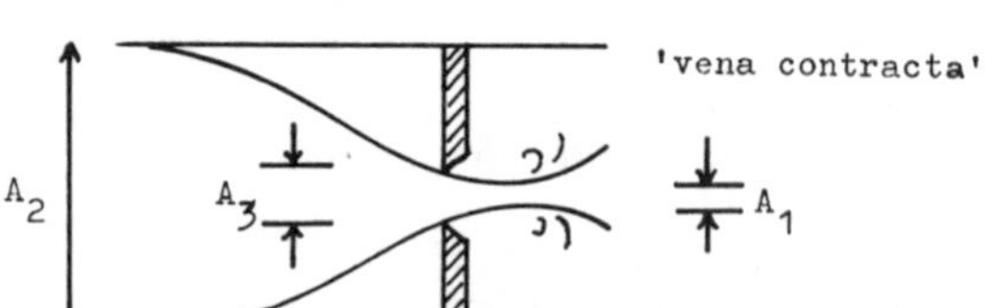

A_1, A_2, A_3 are cross-sectional areas

Rate of discharge, Q

Q = area of flow × velocity of flow

$$Q = (A_3 \times C_c) \times [C_v \sqrt{(2gh)}] = A_3 \times C_D \sqrt{(2gh)}$$

where $C_D = C_c \times C_v$ = *coefficient of discharge.*

Measuring instruments

Pitot tube

Measures velocity head of a flowing liquid.

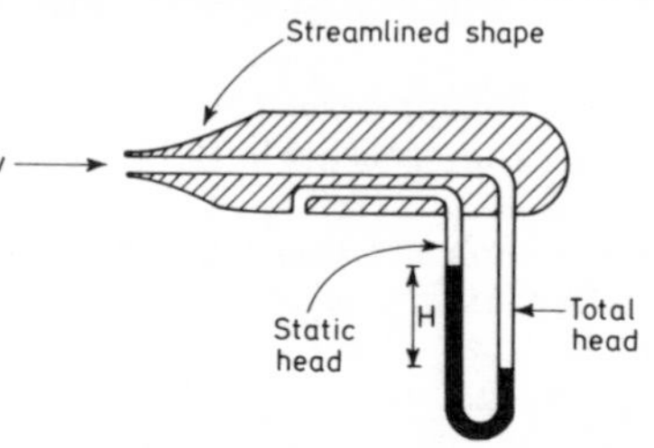

According to Bernoulli Principle:

$$\frac{p_t}{\rho g} + \frac{v^2}{2g} = \frac{p_s}{\rho g}$$

where v = velocity of flow
p_t = total pressure
p_s = static head.

$\therefore v^2 = 2Hg$

For practical Pitot static tube:

$v = C\sqrt{(2Hg)}$

where C is the Pitot-tube coefficient.

Venturi meter

Measures quantity of water flowing through a pipe.

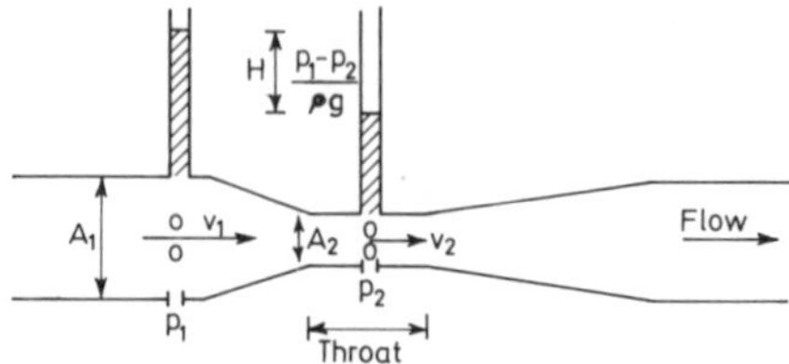

Constricting throat produces increase in velocity and reduction in pressure. For continuity of flow:

$A_1 v_1 = A_2 v_2$

According to Bernoulli Principle (ignoring friction losses):

$$\frac{v_1^2}{2g} + \frac{p_1}{\rho g} = \frac{v_2^2}{2g} + \frac{p_2}{\rho g}$$

$$v_1 = \frac{A_2}{\sqrt{(A_1^2 - A_2^2)}} \sqrt{(2Hg)}$$

For practical discharge measurements:

Actual discharge = $C_D A_1 v_1$

where C_D is a coefficient of discharge.

Gaseous state: a state of matter where atoms or molecules move at random with enormous speed.

GAS LAWS

Expansivity of a gas at constant pressure, α_p

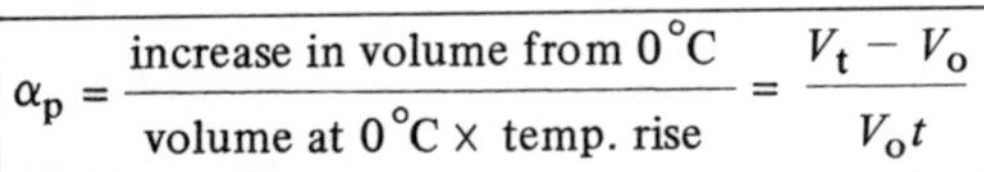

$$\alpha_p = \frac{\text{increase in volume from } 0\,^\circ\text{C}}{\text{volume at } 0\,^\circ\text{C} \times \text{temp. rise}} = \frac{V_t - V_o}{V_o t}$$

for a given mass of gas at constant pressure, or:

$$V_t = V_o\,(1 + \alpha_p t)$$

Charles' Law. A given mass of gas at constant pressure increases by 1/273 of its volume at 0 °C for every 1 °C rise in temperature.

$$V_t = V_o\left(1 + \frac{t}{273}\right)$$

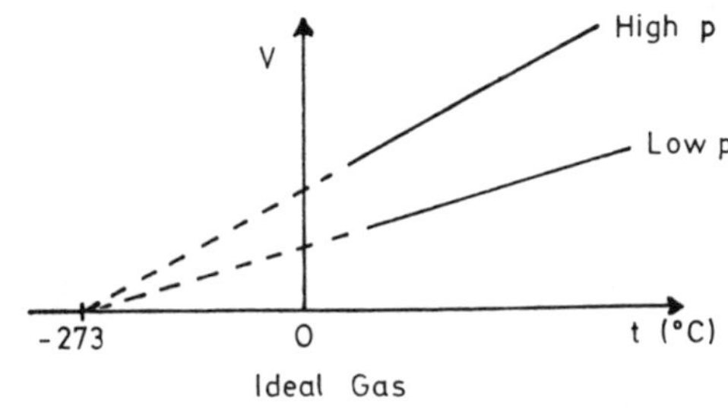

Absolute temperature, T (kelvin, K)

$$T = 273 + t\,^\circ\text{C}$$

Boyle's Law. For a given mass of gas at a constant temperature:

$$pV = \text{constant}$$

where p is the pressure of the gas.

Ideal Gas Law. Combines Charles' and Boyle's Laws:

$$pV = RT \text{ for a } \textit{unit mass} \text{ of gas}$$

where R is a constant
T is the absolute temperature.

$$pV = mRT \text{ for } \textit{any mass} \text{ of gas}$$

where R is the gas constant (per kg)
m is the mass of gas (kg).

$$pV = RT \text{ for a } \textit{mole} \text{ of gas}$$

where R is the molar gas constant.

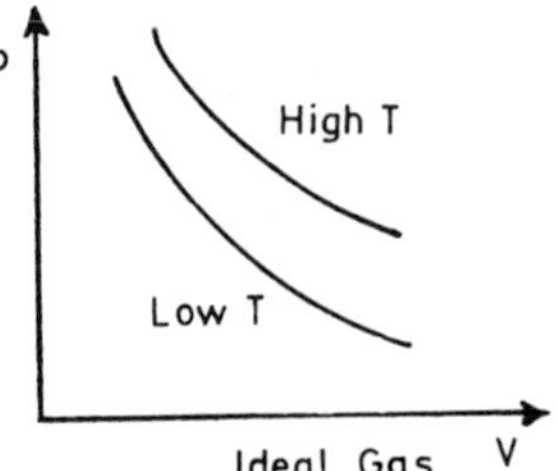

The *mole* is the amount of substance of a system that contains as many elementary entities as there are atoms in 0.012 kg of carbon 12.
Gas constant, R, is determined from the fact that 1 mole of any ideal gas at standard conditions, i.e. at 0°C and 1 atmosphere pressure (1 atm = 101 325 Pa), occupies a volume of 22.413×10^{-3} m^3.

$$\text{Gas constant, } R = pV/T = 8.314 \text{ J K}^{-1} \text{ mol}^{-1}$$

Avogadro's Principle. Equal volumes of all gases at the same temperature and pressure contain equal number of molecules.
At s.t.p. the volume occupied by a mole of any gas is the same ($\approx 22.4 \times 10^{-3}$ m^3).

$$\textit{Avogadro constant, } L = 6.022 \times 10^{23} \text{ mol}^{-1} \text{ at s.t.p.}$$

Boltzmann constant, k

$$k = R/L = 1.381 \times 10^{-23} \text{ J K}^{-1}$$

Dalton's law of partial pressures. The total pressure of a mixture of gases is equal to the sum of the partial pressures that would be exerted by each of the gases occupying a volume equal to that of the mixture.

Real gases. Deviations from the gas laws.
Van der Waals' equation:

$$\left(p + \frac{a}{V^2}\right)(V - b) = RT$$

where a/V^2 is a measure of the attractive forces on the molecules
b is a measure of the dimensions of the molecules.

Reversible change

1. *Isothermal change*: occurs at a constant temperature.

$$pV = \text{constant } (= mRT)$$

2. *Adiabatic change*: no heat enters or leaves the system.

$$pV^\gamma = \text{constant}$$

where γ is the ratio of the specific heat capacities (see page 25).

VAPOURS

Gas: a substance in the gaseous state *above* its critical temperature.

Vapour: a substance in the gaseous state *below* its critical temperature.

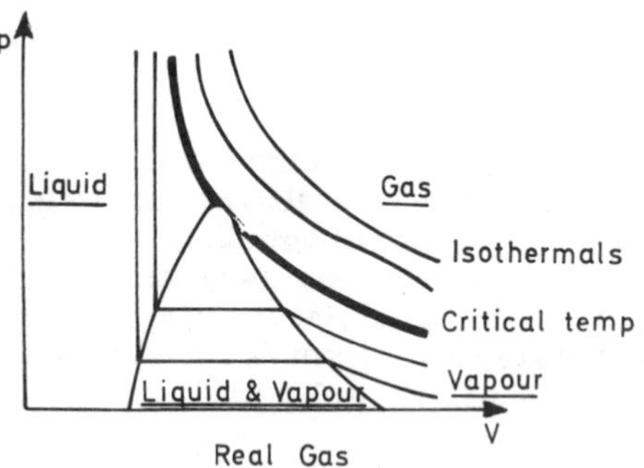

Unsaturated vapour: a vapour *not* in equilibrium with its own liquid.

Saturated vapour: a vapour in equilibrium with its own liquid. The *saturation vapour pressure* (s.v.p.) depends only on the temperature of the liquid. When this pressure equals the external (atmospheric) pressure, boiling occurs.

(a) Constant temperature

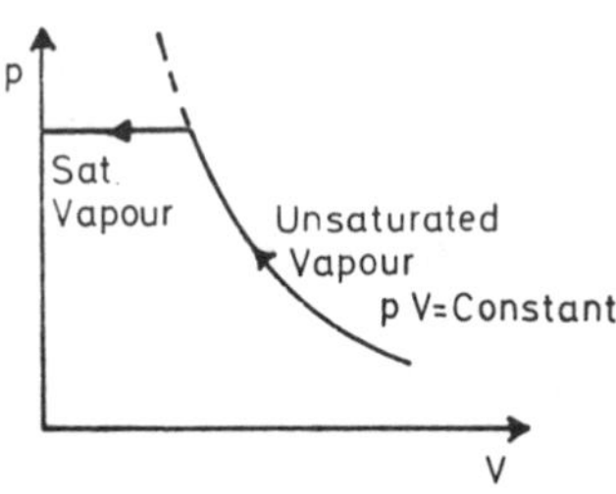

(b) Constant volume

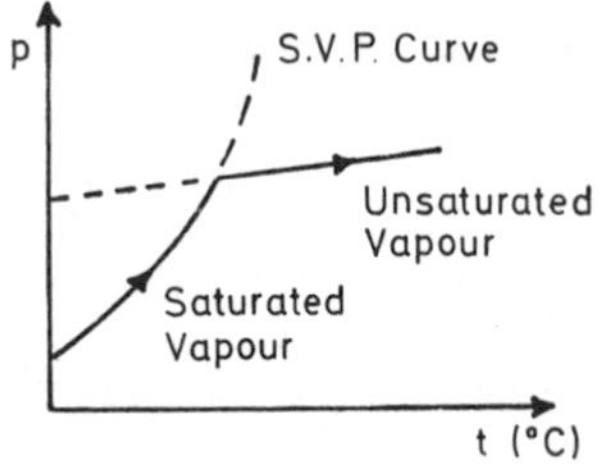

Psychrometry

The study of air and water vapour mixture.

LIQUEFACTION OF GASES

Compression/evaporation refrigeration system

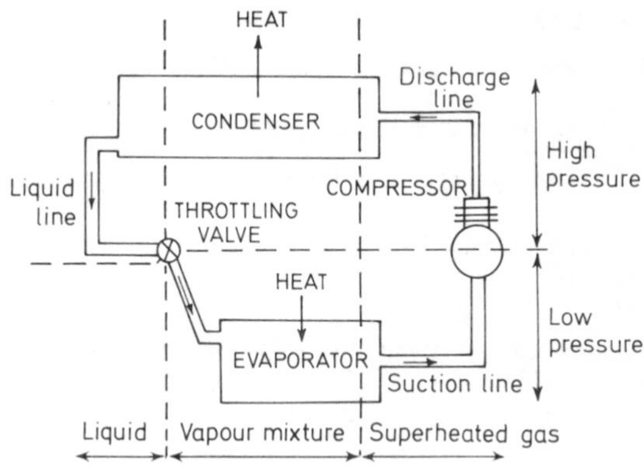

Compression Refrigeration Plant

Properties of refrigerants:

1. Low boiling point, e.g. methyl chloride −24°C, Freon F 12 −29°C.
2. High latent heat of vaporisation.
3. Stable, inert, non-toxic and non-corrosive.
4. Easily compressible.

Heat pump. Compression refrigeration circuit where heat given off at the condenser coil is used as a heat source.

Absorption refrigeration system

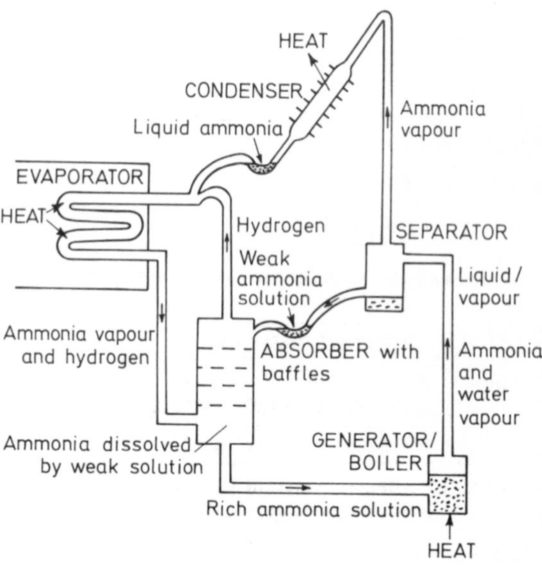

Ammonia Absorption Refrigeration Plant

Charge carriers

In *metals* current is carried by electrons; in *electrolytes*, by ions; in *semiconductors*, by electrons and holes.

Terms

	S.I. units	*Analogy with water*
Electromotive force (e.m.f.), E	volt, V	Static pressure
Potential difference, V	volt, V	Pressure difference
Electrical resistance, R	ohm, Ω	Frictional resistance
Electrical current, I	ampere, A	Rate of flow
Time, t	second, s	

Symbols

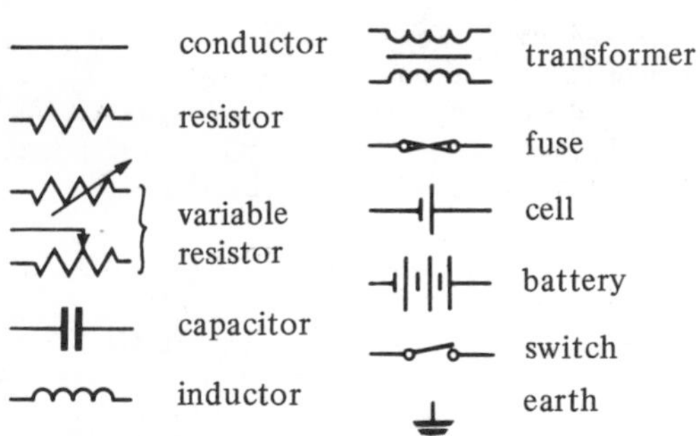

Formulae and units

(a) **Ohm's Law**: for a resistor, $V = IR$ at constant temperature.

(b) **Energy**, E (joules, J):

$$E = VIt$$

The commercial unit of energy is the kilowatt-hour:
1 kW h = 3 600 000 J

(c) **Power**, P (watts, W), is the rate of expenditure of energy:

$$P = VI$$

Heating effect produced by an electric current

Demonstrated by the Joule calorimeter experiment to determine specific heat capacity of water:

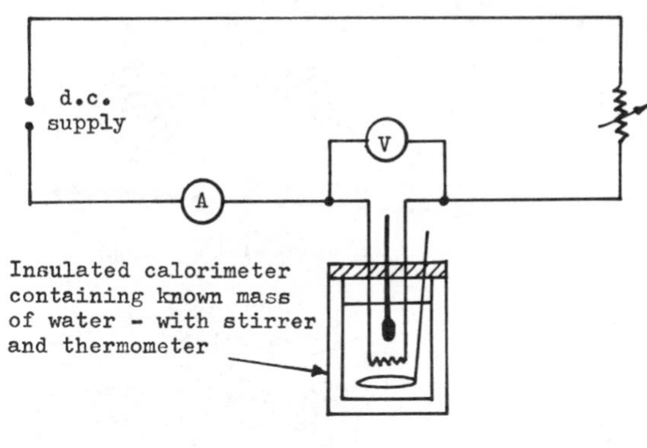

Electrical energy *given to* the system $= VIt$ Joules (1)

where V is the potential difference across the coil (volts)
I is the electrical current through the coil (amps)
t is the time (seconds)

Thermal energy *gained by* the system $= (m_1c_1 + m_2c_2)\theta$ Joules (2)

where m_1 is the mass of the water (kg)
m_2 is the mass of the calorimeter (kg)
c_1 is the specific heat capacity of water (J/(kg K))
c_2 is the specific heat capacity of calorimeter (J/(kg K))
θ is the temperature rise (deg C).

By Principle of Conservation of Energy: equation (1) = equation (2), assuming no heat losses.

Chemical effects produced by an electric current

Two dissimilar metals immersed in an *electrolyte* (conducting liquid) and joined externally cause an electric current to flow, e.g. batteries and electrolytic corrosion.

14 ELECTRICITY (2)

Magnetic effect produced by an electric current

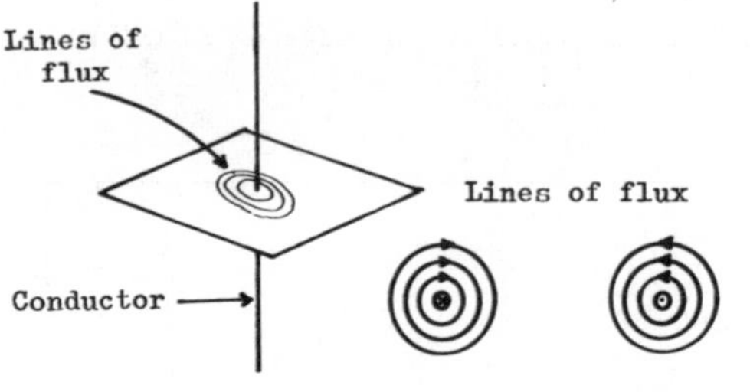

'Cork-screw' rule

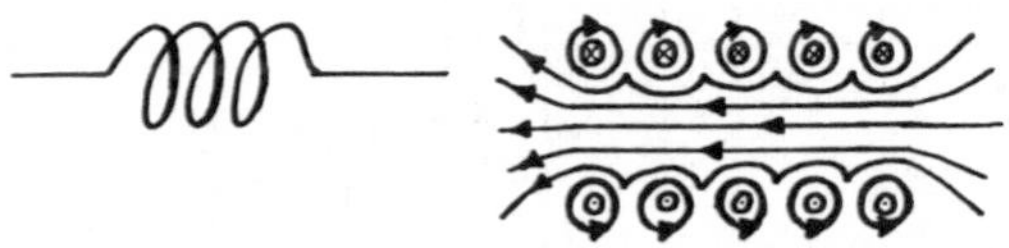

Lines of magnetic flux produced by a solenoid (Coil)

Magnetic field interaction

Magnetic Field

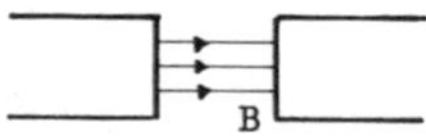

Current Field

Resultant Field

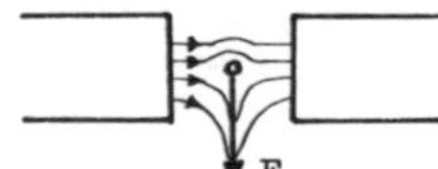

Conductor forced to move from region of high to low flux density

Motors

Convert electrical energy to mechanical energy.

Fleming's left-hand rule for motors:

Magnetic flux (*B*) = index finger
Electric current (*I*) = middle finger
Resultant motion (*F*) = thumb

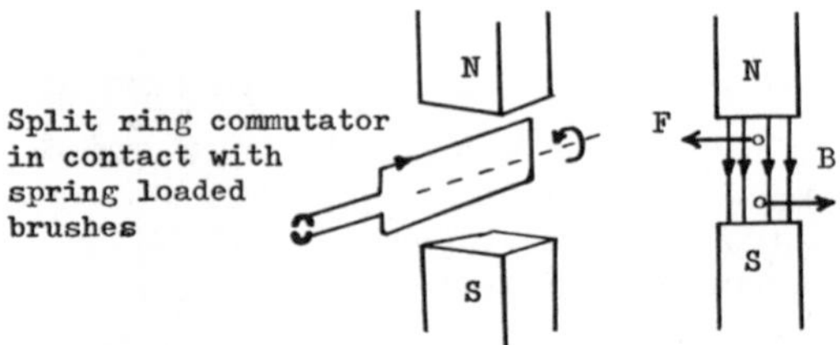

Generators

Convert mechanical energy to electrical energy.

Fleming's right-hand rule for generators:

Magnetic flux (*B*) = index finger
Direction of motion (*F*) = thumb
Induced e.m.f. = middle finger

(a) *alternating current* (a.c.) generation

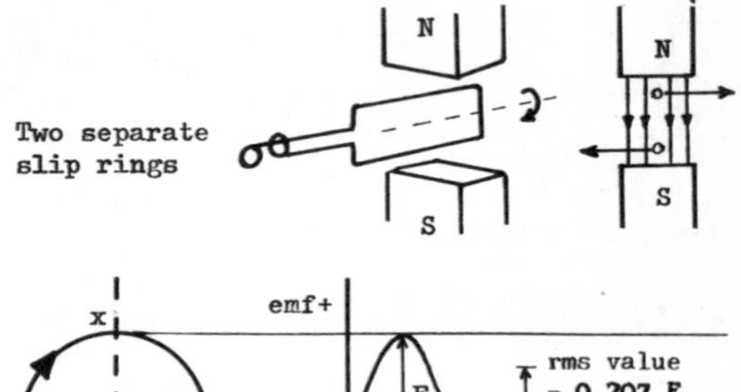

Coil rotating from x to y

Sinusoidal emf waveform generated

The *frequency* of supply is the number of cycles completed per second (units hertz). Practical a.c. generators (*alternators*) invert the position of rotor and stator, and supply 3 phase. Final distributed voltages are usually:

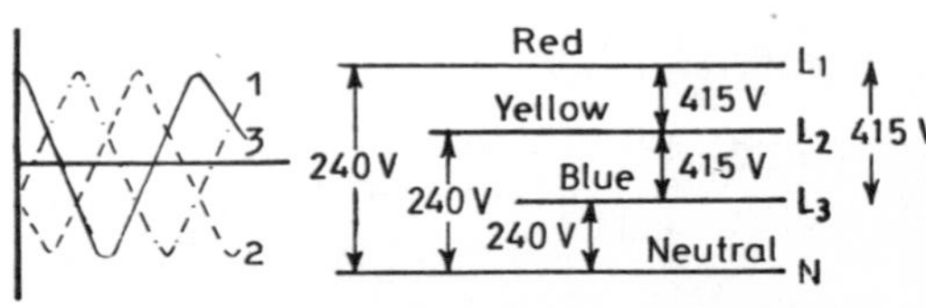

(b) *direct current* (d.c.) generation: *commutator* replaces slip rings on a.c. generator to form a *dynamo*.

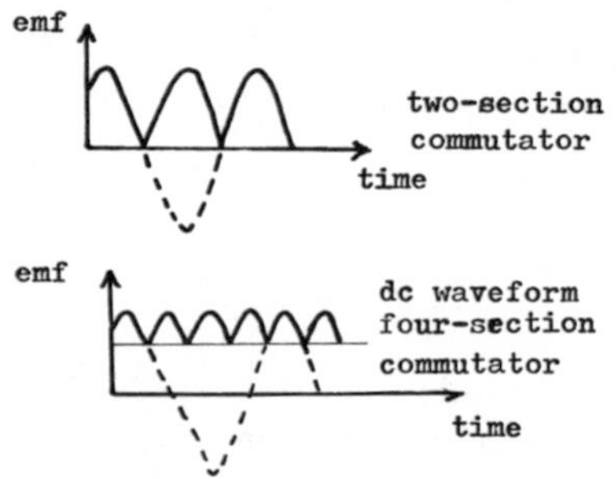

Alternatively *rectifiers* can be used to change a.c. to d.c.

Circuit elements

Resistance, R (ohm)

Resistors are elements that offer physical resistance to current flow and use power.

(a) Resistances in *series*:

$$R_T = R_1 + R_2 + R_3$$

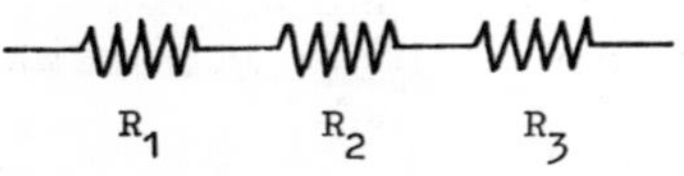

Identical current flows through each resistor. Potential difference across each resistor is proportional to its resistance.

(b) Resistances in *parallel*:

$$\frac{1}{R_T} = \frac{1}{R_1} + \frac{1}{R_2} + \frac{1}{R_3}$$

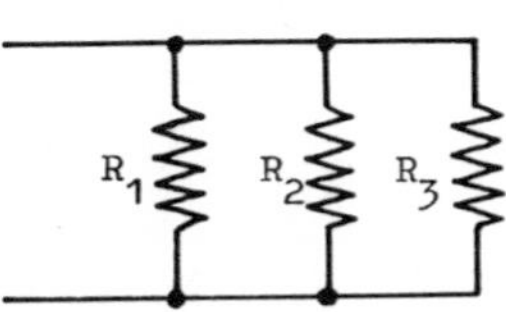

Identical potential difference across each resistor.
Current through each resistor is inversely proportional to its resistance.

Inductance, L (henry)

Inductance occurs when a.c. flows through a solenoid. The changing magnetic flux produced induces a counter current which opposes the initial current. Types: self inductance, mutual inductance.

Transformer: basically two inductances in close proximity. Works only with a.c.

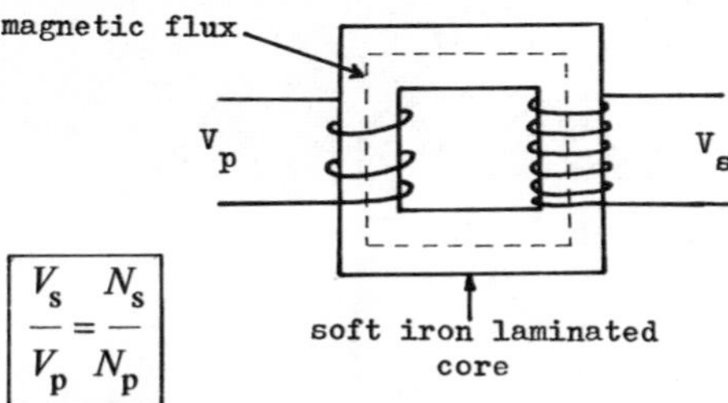

$$\frac{V_s}{V_p} = \frac{N_s}{N_p}$$

where V_p, V_s are primary and secondary voltages
N_p, N_s are numbers of primary and secondary coil turns.

Transformer losses: heat losses
eddy current losses
magnetic flux leakage
hysteresis losses

Capacitance, C (farad)

A device for storing static electricity.
d.c. capacitors act as an open circuit.
a.c. capacitors alternately charge and discharge.

Circuit theorems

Kirchhoff's Laws:

1. The algebraic sum of the currents at any junction of conductors is zero.

$$\Sigma I = 0$$

$$I_1 - I_2 - I_3 + I_4 - I_5 = 0$$

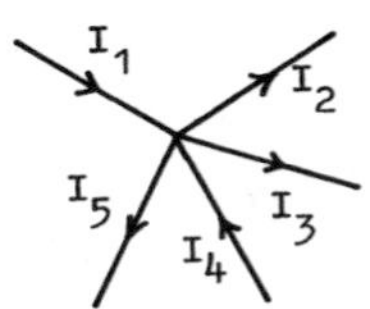

2. The algebraic sum of the voltage drops around a closed circuit loop is zero.
Alternatively:

$$\Sigma(\text{e.m.f.}) = \Sigma(RI \text{ terms})$$

e.g. taking *clockwise* direction as *positive*:

(a) $I_1R_1 - I_2R_2 - I_3R_3 - I_4R_4 = 0$
(b) $I_1R_1 - I_2R_2 - I_3R_3 - I_4R_4 = E$
(c) $I_1R_1 - I_2R_2 - I_3R_3 - I_4R_4 = -E$

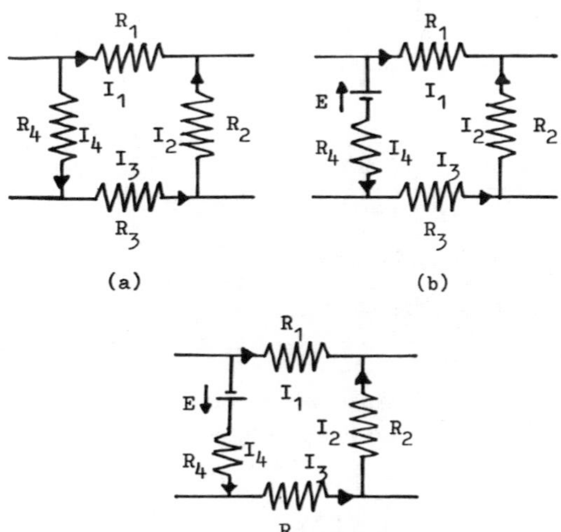

16 ELECTRICITY (4)

A.C. circuits

Impedance, Z (ohm)
Ratio of the root-mean-square (r.m.s.) voltage to r.m.s. current.

$$Z = V/I$$

Admittance, Y (siemens, S)
Reciprocal of impedance.

$$Y = 1/Z$$

A.C. voltage and current

$$I = I_o \cos(\omega t) \qquad V = V_o \cos(\omega t + \phi)$$

where I_o is the maximum current
V_o is the maximum voltage
ϕ is the phase displacement
$\omega = 2\pi f$.

Pure resistance, R

$$V = IR$$

Current and voltage are in phase.

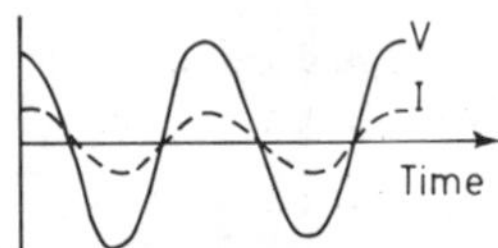

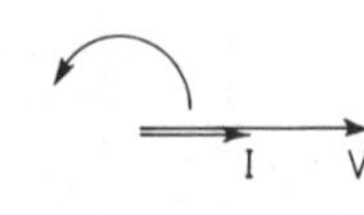

Pure inductance, L

$$V = IX_L$$

where X_L is inductive reactance = ωL.
Current lags voltage. Phase displacement = $\pi/2$.

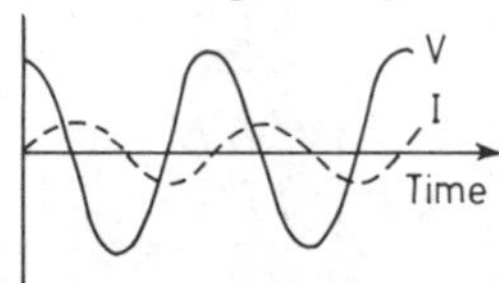

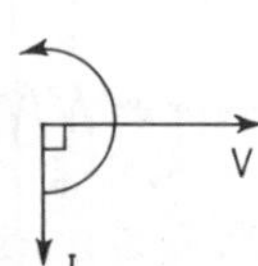

Series L, R circuit:

$$Z = \sqrt{(R^2 + X_L{}^2)}$$

$\phi = \arctan(\omega L/R)$

Pure capacitance, C

$$V = IX_C$$

where X_C is capacitive reactance = $1/\omega C$.
Current leads voltage. Phase displacement = $\pi/2$.

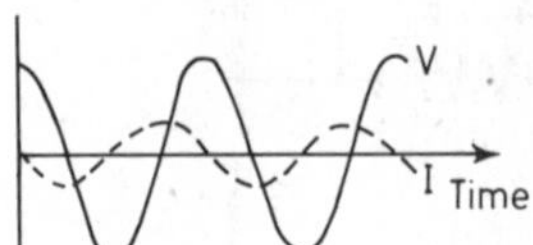

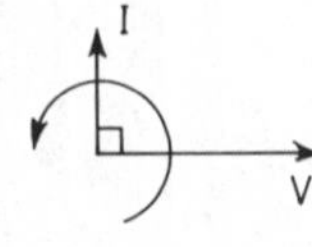

Series C, R circuit:

$$Z = \sqrt{(R^2 + X_C{}^2)}$$

$-\phi = \arctan(1/\omega CR)$

Power, P (watt, W)

$$P = VI \cos\phi$$

Power factor: the ratio of total power (watts) dissipated in circuit to the apparent power (volt-amperes) supplied to the circuit (= $\cos\phi$).

Electrical measurements

1. *Current* measuring instruments:
   galvanometer (d.c. $< 10\ \mu A$)
   moving coil (d.c. up to 5 A)
   moving iron (d.c. and a.c. up to 300 A; Square Law Scale)
   thermoelectric (high-frequency a.c.)
2. *Voltage* measuring instruments:
   current measuring devices with a high resistance in series
   valve voltmeter
3. *Power* measuring instrument:
   wattmeter
4. *Circuit-element* measuring instruments (bridge circuits):
   Wheatstone bridge (resistors)
   Schering bridge (capacitors)
   Owen bridge (inductors)

Wheatstone bridge circuit

At balance:
1. $V_{wx} = V_{wz}$; $V_{xy} = V_{zy}$
2. $\dfrac{R_1}{R_2} = \dfrac{R_3}{R_4}$

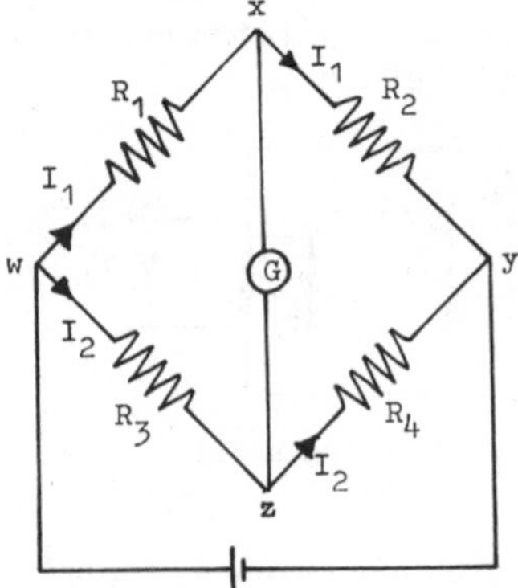

Wheatstone Bridge Circuit

Light: a form of electromagnetic radiation with wavelengths between 380 and 760 mm to which the eye is sensitive. Coherent light waves of the same frequency and phase form LASER beams (Light Amplification by the Stimulated Emission of Radiation).

Light travels in straight lines at a velocity of about 3×10^8 m/s in air. Being transverse waves they can be polarised.

The electromagnetic spectrum

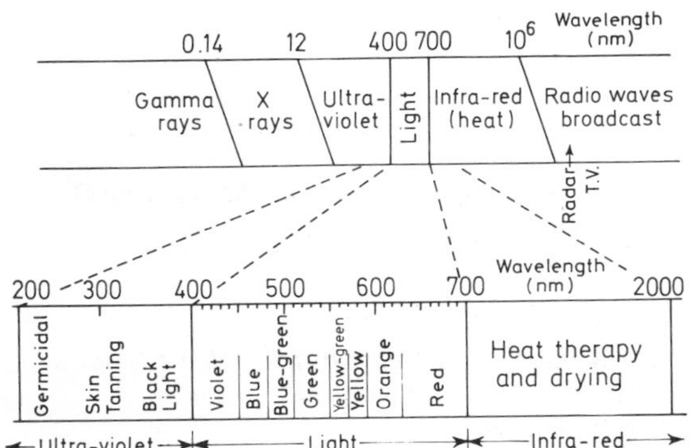

VISION

Visual perception (vision)

Sensory impression produced by the electromagnetic radiation entering the eye.

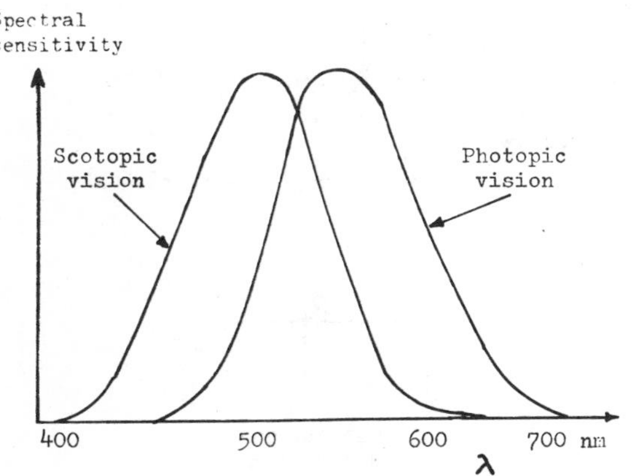

Photopic vision

Vision by the normal eye when it is *light* adapted.
Cone receptors of the retina are the principal active elements.
Spectrum appears *coloured.*

Scotopic vision

Vision by the normal eye when it is *dark* adapted.
Rod receptors of the retina are the principal active elements.
Spectrum appears *uncoloured*, with a shift in the maximum spectral luminous efficiency.

COLOUR

Primary colours: red, green, blue.
Secondary colours: yellow, cyan, magenta.
Methods of *mixing colours*:

1. Additive (coloured lights); 'complementary colours' form *white* light.
2. Subtractive (coloured pigments).

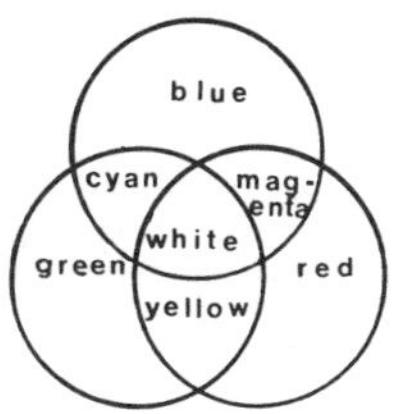

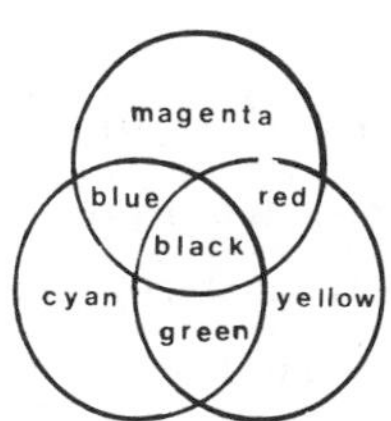

Additive Colour Mixture   Subtractive Colour Mixture

Colour designation: Munsell System or CIE System

Munsell System

Hue	Concept of colour (10 categories).
Value (V)	Subjective measure of reflectance: 0 (absolute black) to 10 (absolute white). Reflectance = $V(V - 1)/100$.
Chroma	Intensity of colour (16 classes). A low chroma denotes a pastel shade.
Greyness	Not always specified.

Example designation: 7.5 BG 6/4.

CIE Standard Colorimetric System
Colorimetric system for evaluating any spectral distribution of the power with the aid of three functions of wavelength, the CIE spectral tristimulus values: $\bar{x}(\lambda)$, $\bar{y}(\lambda)$, $\bar{z}(\lambda)$.

18 LIGHT (2)

Colour temperature (kelvin, K): the temperature of a *black body* that emits radiation having a chromaticity nearest to that of the light source being considered.

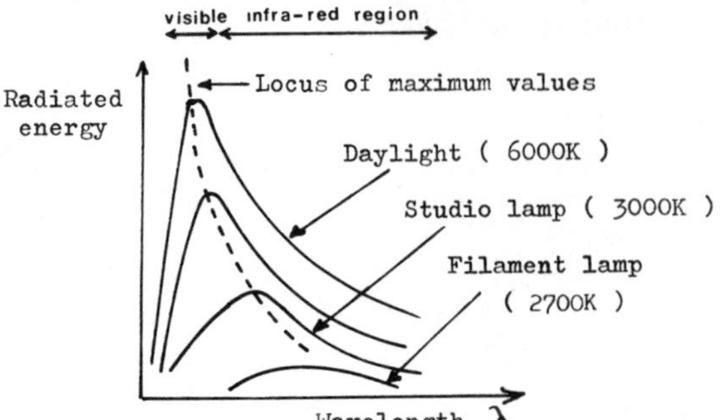

Colour appearance: the apparent colour of light emitted from a source.

Colour rendering: the colour of objects seen in the light from a source.

Colour rendering index, R: a measure of the degree to which measured colours of objects conform to those of the same objects under a reference illuminant.

GEOMETRICAL OPTICS

Transmission of light

Materials may be opaque, transparent or translucent.

Reflection of light

A matt surface causes diffuse reflection (e.g. white blotting paper has a reflectance of 85%).

A plane surface causes specular reflection (e.g. mirror).

Diffuse reflection

Specular reflection

Partial specular reflection

Laws of specular reflection

1. The incident ray, the reflected ray and the normal all lie in the same plane.
2. The angle of incidence equals the angle of reflection, measured to the normal.

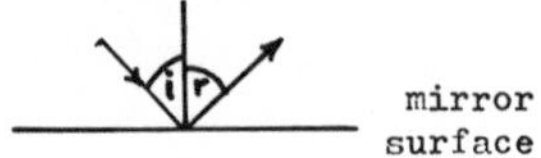

Refraction of light

Occurs when light passes through a surface between two media (e.g. air/glass).

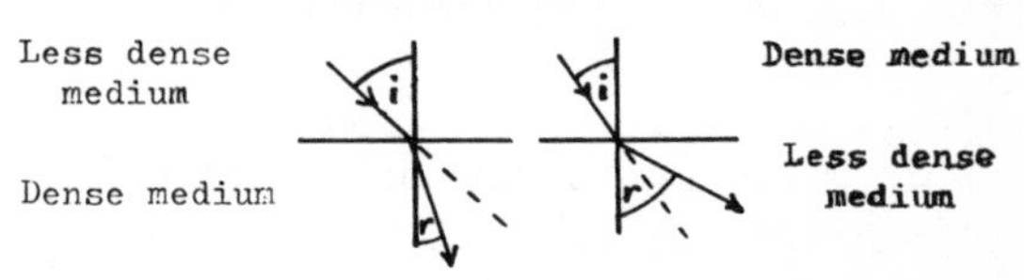

Refractive index, n

A measure of the optical density of a medium.

$$n = \frac{c}{c_n}$$

where c is the speed of monochromatic light in vacuo

c_n is the phase velocity of the waves in the medium.

Laws of refraction

1. The incident ray, the normal and the refracted ray all lie in the same plane.
2. **Snell's Law:**

$$n_1 \sin i = n_2 \sin r$$

where i is angle of incidence in medium 1

r is angle of refraction in medium 2

n_1 and n_2 are refractive indices of media 1 and 2.

Total internal reflection occurs when:

(a) light travels from a dense to a less dense medium,

(b) the angle of incidence is greater than the critical angle.

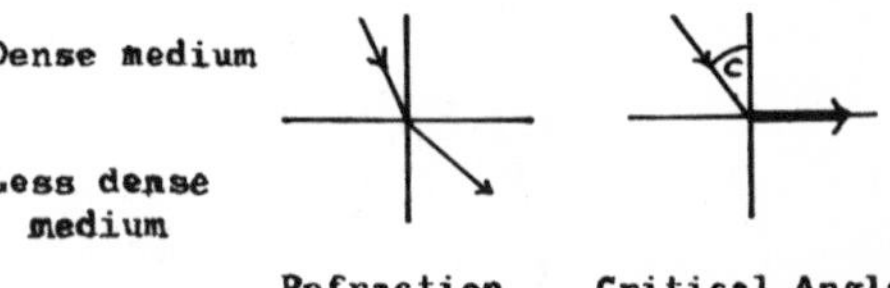

Refraction   Critical Angle

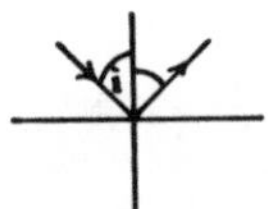

Total Internal Reflection

Refraction of white light through a prism

Dispersion occurs to form the white-light visible spectrum (ROYGBIV).

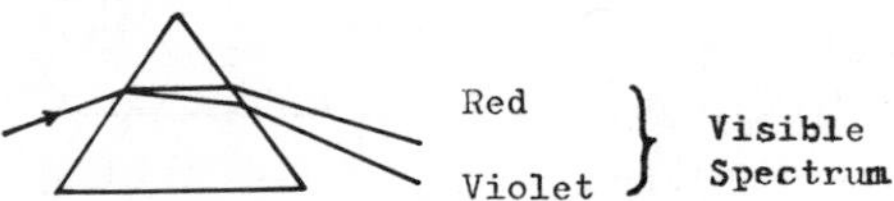

Refraction of light through lenses

Types of lens: *Convex*, or converging
Concave, or diverging

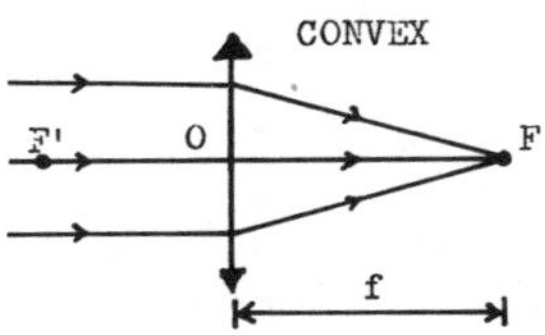

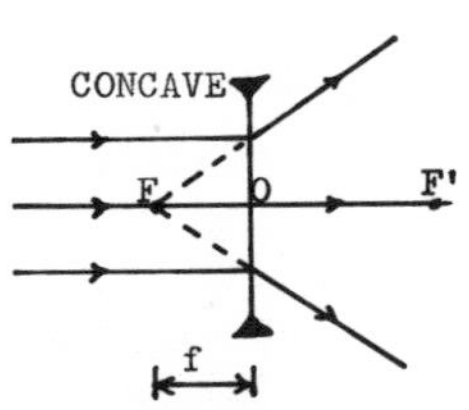

F = Principal Focus
OF = Principal Axis
f = Focal Length (2 per lens)
O = Optical Centre

Lens formulae:

$$\frac{1}{f} = \frac{1}{u} + \frac{1}{v}$$

$$\text{Magnification} = -\frac{v}{u}$$

where u = distance of object from lens
v = distance of image from lens
f = focal length of lens

Surveyor's telescope

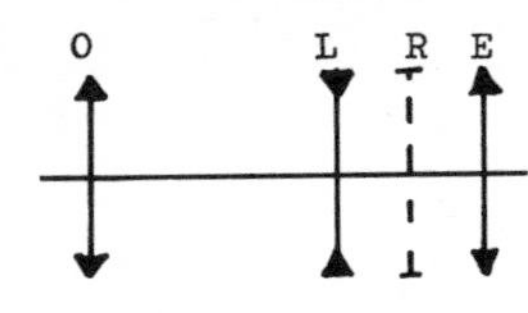

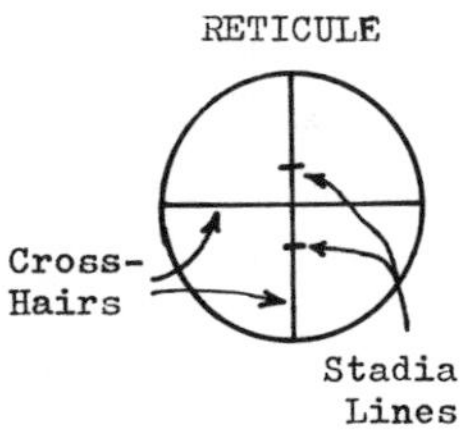

O = Objective Lens
L = Focussing Lens
R = Reticule
E = Eyepiece

ARTIFICIAL LIGHTING

Units of light

Intensity, I (candela, cd)
A measure of the illuminating power of a light-source in a particular direction, independent of the distance from the source.

Luminous flux, Φ (lumen, lm)
The 'flow' of light. One lumen of flux is emitted in one *steradian*, by a point source of uniform luminous intensity of one candela.

Illuminance, E (lux, lx)
The luminous flux incident per unit area of surface.

$$E = \Phi/A$$

where A is the area (m^2).

Luminance, L ($cd\ m^{-2}$)
The luminous flux emitted per unit area of surface in a given direction.
Note. 1 apostilb = $1/\pi$ $cd\ m^{-2}$.
Luminance = 0.318 × illuminance × reflectance.

Luminosity
Visual sensation associated with luminance.

Light sources

Incandescent lamps

A lamp in which light is produced by heating a filament.

(a) *Tungsten filament (GLS) lamps*: point sources; low capital cost; large quantity of heat produced; efficacy about 13 lm/watt; lamp life 1000 hours; easily dimmed.

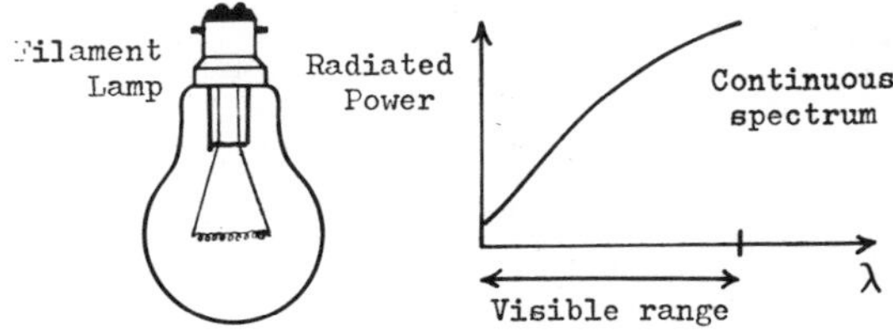

(b) *Tungsten halogen (TH) lamps*: usually pencil-shaped sources in special fittings; operate at higher temperature than GLS lamps, hence quartz envelope; efficacy about 24 lm/watt; lamp life 2000 hours.
(c) *Reflector (PAR) lamps*: tungsten filament lamps with internal reflectors.

20 LIGHT (4)

Discharge lamps

A lamp in which light is produced by an electric discharge through a gas, a metal vapour or a mixture of both.

(a) *Sodium lamps*: low-pressure (SOX, SLI/H) lamps produce mainly yellow light with efficacy of 180 lm/watt; lamp life 4000 hours. High-pressure (SON, SON/T) lamps with better colour rendering are also made.

(b) *Mercury lamps*: produce a discontinuous light spectrum with virtually no red light.
Non-corrected mercury discharge lamps (MB): high-pressure lamps for specialist applications, providing whiter light than a low-pressure lamp.
Colour-corrected mercury lamps (MBF, MBFR): high-pressure lamps with phosphor coating on inside of glass envelope (outer bulb) to improve colour-rendering properties.
Blended lamps (MBT, MBTL): high-pressure mercury discharge lamps with tungsten ballast (no external control gear required).
Mercury halide lamps (MBI): colour of light output improved by addition of halides.
Tubular fluorescent lamps (MCF): linear sources with control gear; lower radiated heat; nominal lamp life 5000–7000 hours but lamps subject to aging; flicker and stroboscopic effects; luminous efficacy (30–60 lm/watt) and colour-rendering properties dependent on phosphors.

Tubular fluorescent lamp

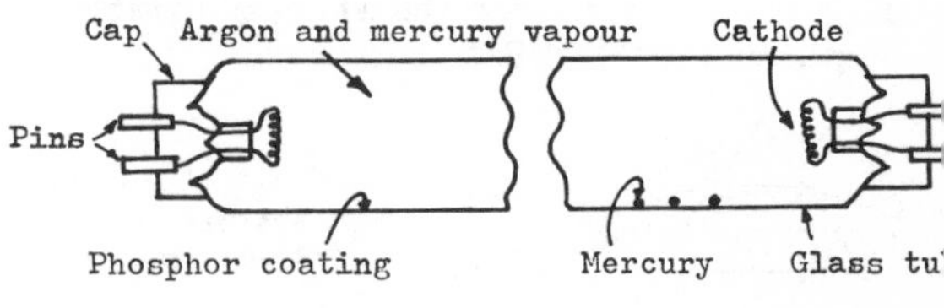

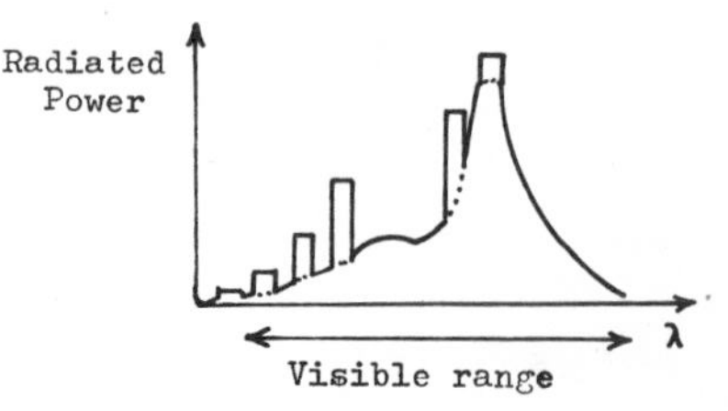

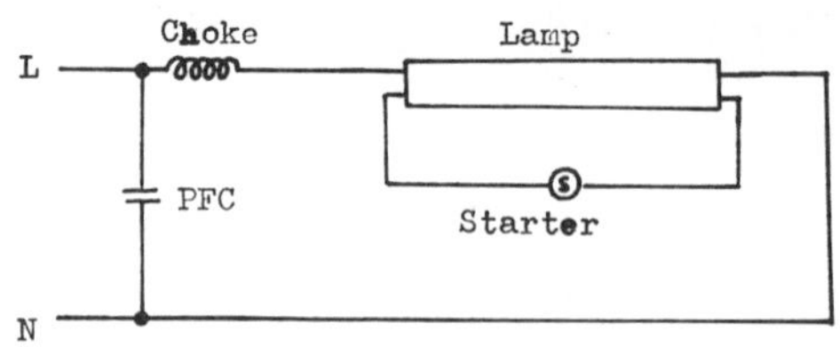

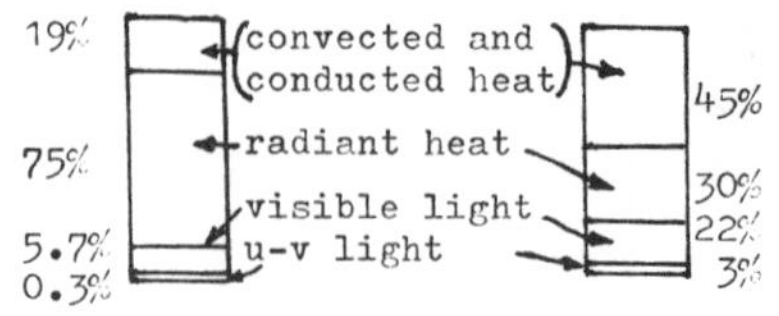

Properties of various types of fluorescent lamps

Colour temperature	*Source of same colour range*	*Fluorescent tube and relative efficacy*		
		High-efficacy (100–95)	*De-luxe* (75–65)	*Special* (65–40)
12 000 K	Clear sky			
6 500 K	Overcast sky		*Northlight/colour matching*	Artificial daylight
4 000 K	Afternoon sun	*Daylight*	*Natural*	Trucolor 37
3 500 K		*White*		
3 000 K	Filament lamp	*Warm white*	De-luxe warm white	Softone 27
2 000 K	Candle			
CIE colour rendering index:		50–70	70–95	95–100

Lamp names in italics are standard designations (BS 1853).

Lighting design for building interiors

Standards and legislation
IES Code 1977, for interior lighting.
Theatres Act, 1968.
Shops, Offices and Railway Premises Act, 1963.
Health and Safety at Work Act, 1974.
Building Regulations, 1976.
SR & O No. 731: Chemical Works Regulations, 1922.
SI No. 2168: Slaughterhouse (Hygiene) Regulations, 1958.
SI No. 1172: Food Hygiene (General) Regulations, 1970.
SI No. 890: Standards for School Premises Regulations, 1959 amended 1970.
IEE Regulations.
British Standards and Codes of Practice

Quantity

IES recommended standard service illuminance

Task/interior	*Standard Service Illuminance* (lux)
Storage areas	150
Casual work	200
Rough work	300
Routine work	500
Demanding work	750
Fine work	1000
Very fine work	1500
Minute work	3000

(a) **Point sources: Cosine Law of Illumination**

$$E_p = \frac{I}{D^2} \cos\theta$$

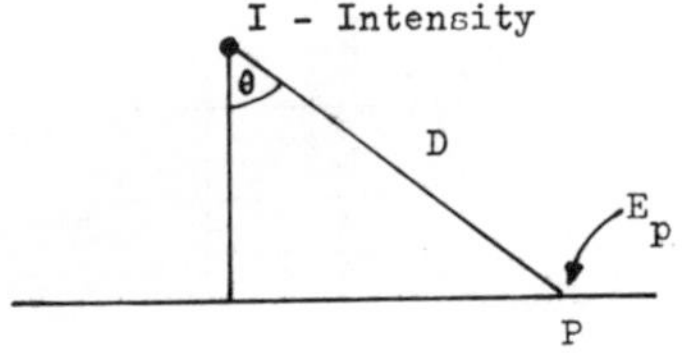

General case:

$$E_\gamma = \frac{I_\theta \cos^2\theta \cos\gamma}{h^2} \text{ lux}$$

where E_γ is the illuminance at P on any plane the normal to which makes an angle γ with the direction of incidence of the light.

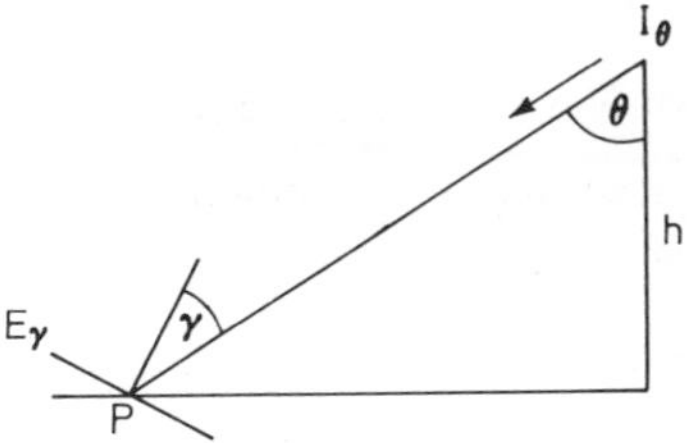

(b) **Regular array of luminaires: Lumen Method**

$$E = \frac{nFUM}{A}$$

where E = average illuminance on horizontal working plane (lux)
n = number of luminaires
F = total flux emitted per luminaire (lumens)
M = maintenance factor
A = area illuminated (m^2)
U = utilisation factor (obtained from tables), which depends on:
(a) room proportions defined by *room index*
(b) absorption of light in luminaire
(c) absorption of light at various room surfaces.

$$\textbf{Room index} = \frac{LB}{H_m(L+B)}$$

where L and B are length and breadth of room
H_m is the mounting height (working plane normally 0.85 m).
Maximum spacing of luminaires = $1.5 \times H_m$.

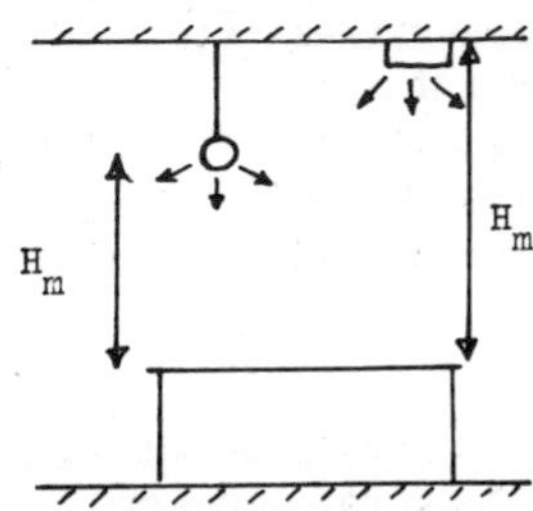

(c) Classification of luminaires

Light Output Ratio (LOR)

Classification of luminaires according to the proportion of the total light output of the luminaire in the upper and lower hemispheres.

$$\text{LOR} = \frac{\text{Total light output of luminaire}}{\text{Light output of its lamp(s)}}$$

Downward LOR (DLOR) =

$$\frac{\text{Light output of luminaire downwards}}{\text{Light output of its lamp(s)}}$$

Upward LOR (ULOR) =

$$\frac{\text{Light output of luminaire upwards}}{\text{Light output of its lamp(s)}}$$

LOR = DLOR + ULOR

e.g. LOR = (300 + 450)/1000 = 75%
DLOR = 450/1000 = 45%
ULOR = 300/1000 = 30%

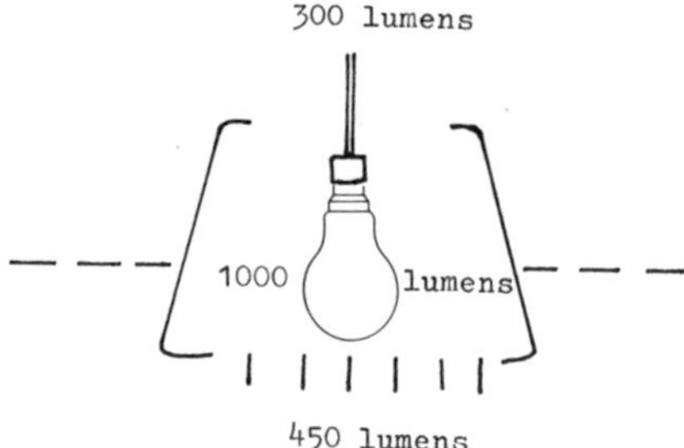

British Zonal (BZ) system

Ten polar curves for classifying luminaires according to their downward light distribution: BZ1 mainly downward, BZ10 mainly upward. Most luminaire curves cross from one BZ reference zone to another depending on room index; e.g. BZ5/2/BZ4 indicates a luminaire classified BZ5 for room indices up to 2, and BZ4 for indices greater than 2.

The *direct ratio* is the proportion of the total downward flux from an installation of luminaires that is directly incident on the working plane.

Special luminaires

Proof luminaires (flame-, dust-, vapour-, drip-, rain-, jet-proof). Air-handling luminaires.

Polar curves in the BZ classification. Curves relate only to lower hemisphere and are scaled to 1000 lumens.

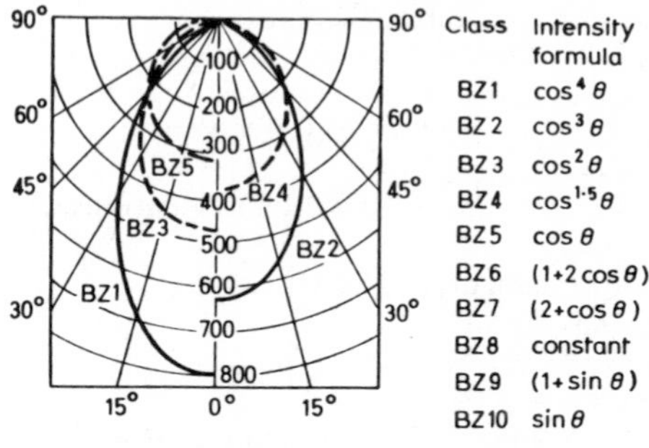

Class	Intensity formula
BZ1	$\cos^4\theta$
BZ2	$\cos^3\theta$
BZ3	$\cos^2\theta$
BZ4	$\cos^{1.5}\theta$
BZ5	$\cos\theta$
BZ6	$(1+2\cos\theta)$
BZ7	$(2+\cos\theta)$
BZ8	constant
BZ9	$(1+\sin\theta)$
BZ10	$\sin\theta$

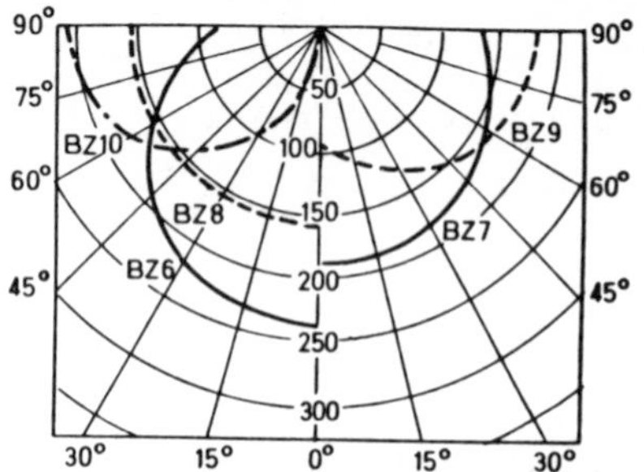

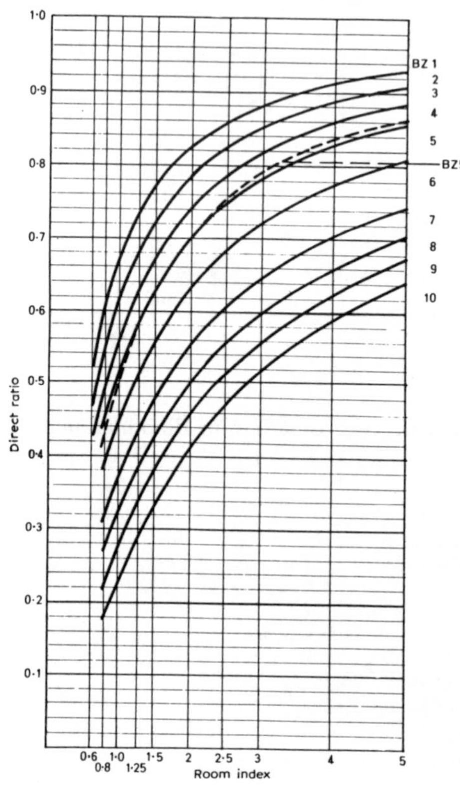

Quality

(a) **Types of lighting**
General, localised, local.
Direct, indirect, directional, diffused.
Emergency, escape, standby, extra-low voltage (below 50 V).

(b) **Glare**
Types: *discomfort* and *disability* glare.

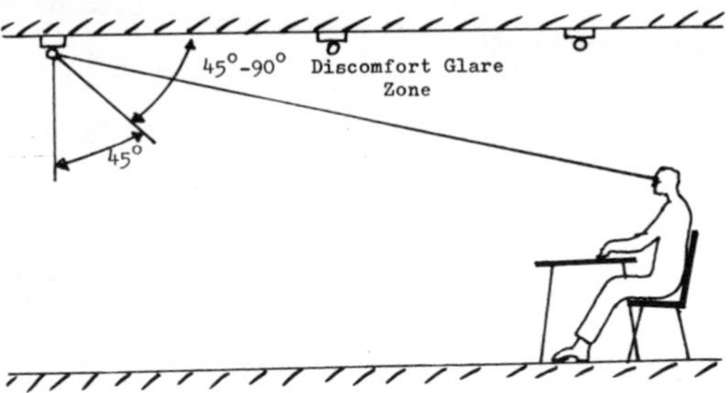

Glare constant for a single luminaire, g:

$$g = K \frac{L_s^{1.6}\ \omega^{0.8}}{L_b} \cdot \frac{1}{p^{1.6}}$$

where L_s = luminance of source (cd m^{-2})
L_b = background luminance (cd m^{-2})
p = position index
ω = solid angle subtended by source at the eye (steradians)
K = constant = 0.478.

Glare constant for an installation, G:

$$G = g_1 + g_2 + \ldots + g_n$$

Glare index, GI:

$$\text{GI} = 10 \log (G \times 0.5)$$

Glare indices usually range from 10 to 30 in increments of 3 units.

Low-brightness luminaires: 10, 13, 16.
Bare lamps in large open rooms: 28, 30.
Recommended *limiting glare index* for offices: 19.

(c) **Illuminance and Modelling**

Planar illuminance, E (lux)
Measurement of illuminance on a plane.
Measurement instrument: cosine, colour-corrected photocell.

Scalar (or mean spherical) illuminance, E_s (lux)
Average illuminance, at a point, on the surface of an infinitesimally small sphere at that point.
Measurement instrument: two photocells placed back to back and covered by small translucent diffusing spheres.

Illumination vector, $\vec{E}$ or E (lux, and direction)
Term used to describe the directional characteristics of light at a point.

Measurement instrument: illumination vector photometer.

Vector/scalar ratio
Measure of the directional strength of light.
Ratio of the illumination vector to the scalar illuminance.
Values between 1.2 and 1.8 produce a pleasing effect.

Modelling
Faces have a pleasing appearance when seen in light with a vector direction of between 15° and 45° below the horizontal.

(d) **Flicker and stroboscopic effect**

Flicker: impression of fluctuating luminance or colour, occurring when the frequency of the variation of the light stimulus lies between a few hertz and the *fusion frequency* of the retinal images.

Stroboscopic effect: apparent change of motion or immobilisation of an object when the object is illuminated by a periodically varying light of appropriate frequency. Can be eliminated by supplying adjacent luminaires on different phases, or using twin-lamp fittings on *lead-lag* circuits, or providing 'local' lighting from tungsten-filament lamps.

NATURAL LIGHTING

Sky luminance distributions

1. Clear sky luminance (cloudless day)
2. Uniform sky luminance
3. CIE Standard Overcast Sky

CIE Standard Overcast Sky

Used as the basis for daylight predictions in most temperate climates. Assumes *no* direct sunlight reaches the ground.

$$L_\theta = \frac{L_z}{3}(1 + 2 \sin \theta)$$

where L_z = sky luminance at the zenith
L_θ = sky luminance at an altitude of θ degrees above the horizon.

Illuminance due to unobstructed hemisphere of overcast sky is assumed to be 5000 lux (for calculations).

Daylight factor (%)

The instantaneous illuminance (E_i) at a point inside a building, expressed as a percentage of the simultaneous horizontal illuminance (E_o) received from an unobstructed hemisphere of sky (no direct sunlight).

Daylight factor = $(E_i/E_o) \times 100$
Daylight factor = SC + ERC + IRC

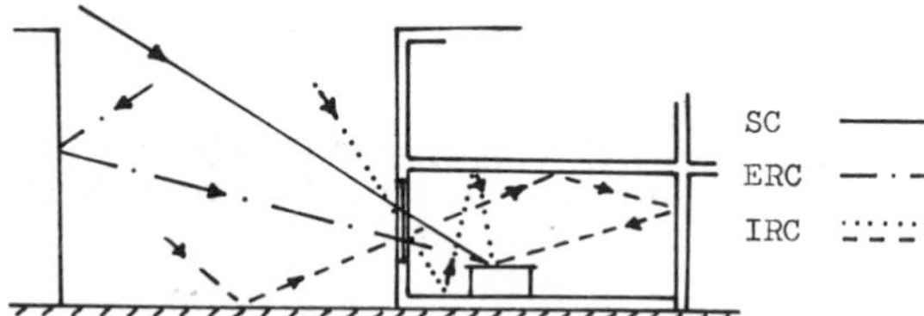

1. *Sky Component (SC) of daylight factor (%).* The illuminance received at an internal reference point directly from the sky, expressed as a percentage of the simultaneous illuminance outside.
2. *Externally Reflected Component (ERC) of daylight factor (%).* The illuminance received at an internal reference point from external reflecting surfaces, expressed as a percentage of the simultaneous illuminance outside.
3. *Internally Reflected Component (IRC) of daylight factor (%).* The illuminance received at an internal reference point from internal reflecting surfaces, expressed as a percentage of the simultaneous illuminance outside.

Prediction methods

1. *SC & ERC* (Reference BRS Digest 41):
   (a) BRS daylight protractors
   (b) Waldram diagrams
   (c) perspective methods
   (d) simplified daylight tables
   (e) equations
2. *IRC* (Reference BRS Digest 42):
   (a) IRC tables
   (b) nomograms
   (c) BRS inter-reflection formula
3. Additional corrections for:
   (a) deterioration of decorations (IRC only)
   (b) dirt on glass
   (c) window framing and bars
   (d) alternative types of glazing materials

Measurement

1. Artificial skies for model studies:
   (a) rectilinear mirrored type
   (b) domical-types
2. Daylight factor meters for room studies.

Design for daylight in buildings

PSALI (Permanent supplementary artificial lighting installation)

$E = \pi DL/10$

where E = level of supplementary artificial lighting (lux)
L = sky luminance (cd m^{-2})
D = average daylight factor over the area supplemented (%).

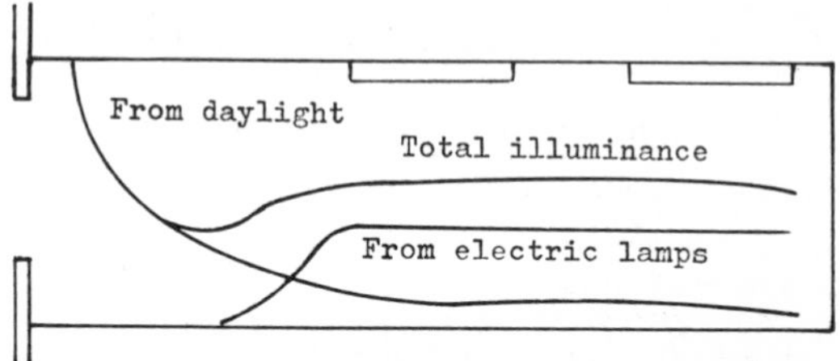

Total illuminance in an interior with side windows

IED (Integrated Environmental Design)
A concept in building design that aims to maintain an optimal internal environment for comfort, efficiency and low energy consumption throughout the year, irrespective of the external climate. The whole structure and services are designed to minimise heat losses and heat gains, by making maximum use of heat recovery and the cost of the engineering services required to control the internal temperature, humidity, freshness, air cleanliness and lighting standards.

Temperature: the condition of a body that determines whether heat shall flow from it.

Scales: thermodynamic temperature, T (kelvin, K)
common temperature, t (Celsius, °C)

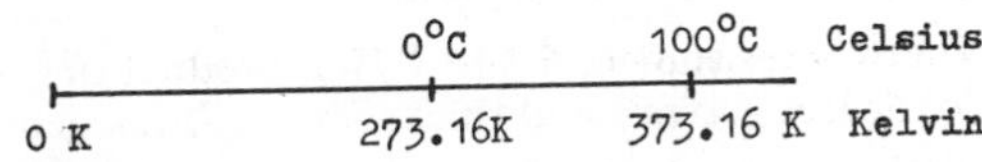

Heat, Q (joule, J): a form of energy. The amount of thermal energy (Q) an object retains depends on:
(a) the temperature of the object,
(b) the mass of the object,
(c) the *heat capacity* of the object.

Heat capacity, C (J K^{-1}): product of mass (m) and *specific heat capacity* (c).

$$C = mc$$

Specific heat capacity, c (J kg^{-1} K^{-1}): the heat energy received or emitted from a unit mass of substance when heated or cooled by 1 kelvin *without a change of state.*

$$c = \mathrm{d}Q/\mathrm{d}T$$

Gases have *two* specific heat capacities:
(a) specific heat capacity at constant pressure, c_p
(b) specific heat capacity at constant volume, c_v

$$\text{Ratio of the specific heat capacities} = \gamma = c_p/c_v$$

Specific latent heat, l (J kg^{-1}): the heat energy required to change a unit mass of substance from one state of matter to another *without a change of temperature.*

Two types: specific latent heat of fusion
specific latent heat of vaporisation (evaporation)

Effects of heat on matter

(a) *Either* a rise in temperature:

$$\text{Sensible heat} = mc\theta \text{ J}$$

where θ is the temperature rise/fall (K);
and/or a change of state.

(b) Increase in dimension. Types of *coefficients of expansivity*:
For *solids*: linear, superficial, cubic.
For *liquids*: apparent, absolute (or real).
For *gases*: cubic, at constant pressure or constant volume (see section on Gases).

Coefficient of linear expansivity, α (K^{-1}): the fractional change in length ($\Delta L/L_o$) per unit rise in temperature (θ).

$$\alpha = \Delta L/L_o\theta$$

where L_o is the original length.

(c) Increase in electrical resistance of pure metals. Decrease in electrical resistance of carbon and non-metals.

(d) Heat applied to the junction of two dissimilar metals induces a small e.m.f.—a *thermocouple.* (e.g. copper/constantan)

Methods of heat transfer

Conduction

(a) Molecular vibrations increase with body temperature.
(b) Energy is transferred to 'touching' molecules.
(c) Conduction effects are most noticeable in solids.

Convection

(a) Occurs in fluids.
(b) Heat applied to fluid causes expansion; hence reduces density and allows molecules to rise, only to be replaced by colder, more dense molecules.

Radiation

(a) Electromagnetic waves that travel in straight lines but can be reflected.
(b) Does *not* require a medium for transmission.
(c) Radiant heat more dominant with increasing body temperature.

Radiant exitance, M (W m^{-2})
From a *black body* (perfect absorber):

$$\textit{Stefan-Boltzmann Law: } M_b = \sigma T^4$$

where M_b is radiant exitance from a black body (W m^{-2})
σ is the Stefan-Boltzmann constant $= 5.7 \times 10^{-8}$ W m^{-2} K^{-4}
T is the absolute temperature (K).

$$\textit{Wien's Displacement Law: } \lambda_{max} T = \text{constant}$$

where λ_{max} is the dominant wavelength (m)
Constant $= 2.898 \times 10^{-3}$ m K.

From a *real body*:

$$\epsilon = M/M_b$$

where ϵ is emissivity
M is radiant exitance from the real body at the same temperature as the exitance M_b from a black body radiator.

Surface emissivity, ϵ

Low emissivity
Brightly polished surface (e.g. polished aluminium, 0.05).
Relatively high surface temperature.
Tends to prevent heat loss from the surface.

High emissivity
Dull black surface (e.g. dark brick, 0.9).
Relatively cool surface temperature.
Provides very little insulating efficiency to structure.

Newton's Law of Cooling. The rate at which a body loses heat is proportional to the excess temperature above ambient conditions. (Strictly, applies only to forced convection.)

HEAT LOSSES FROM BUILDINGS

Ventilation heat losses, Q_v (watt, W)

$$Q = mc_p\theta$$

where m = mass of air (kg)
c_p = specific heat capacity of air at constant pressure (= 1.012 kJ kg^{-1} K^{-1})
θ = internal/external design air temperature difference (K)

But mass (m) = density (ρ) × volume (V)

$$Q_v = V_t \rho c_p \theta = \frac{(1.205)(1012)NV\theta}{3600} = 0.33\ NV\theta$$

where V_t = total volume change of air per second (m^3)
V = volume of the enclosure (m^3)
N = number of air changes per hour
ρ = density of air (= 1.205 kg m^{-3})

Heat loss through fabric, Q_c (watt, W): depends on type of material, form of construction, degree of exposure, and orientation. The calculations below assume 'steady state' or 'equilibrium' conditions.

Fabric heat loss calculations

Thermal conductivity, k (W/(m K)) of a solid material is the quantity of heat conducted through a cross-sectional area of 1 m^2 to depth of 1 m in one second, if the temperature differential between the faces is maintained at 1 deg C.

$$\text{Rate of flow of heat} = dQ/dt = -kA(d\theta/dx)$$

where $d\theta/dx$ is the temperature gradient.
Measured by *Searle's Apparatus* for good conductors and by *Guarded Hot Plate* for bad conductors.

Thermal resistivity, r (m K/W): reciprocal of thermal conductivity.

$$r = 1/k$$

Thermal conductance, C (W/(m^2 K)): the surface-to-surface conductance of a material or construction.

$$C = k/L$$

where L = thickness in metres.

Thermal resistance, R (m^2 K/W): product of thermal resistivity and thickness of a homogeneous material.

$$R = rL = L/k$$

Cavity resistances: unventilated cavities greater than 20 mm wide have a typical resistance of 0.18 m^2 K/W. Thermal resistance can be increased by placing aluminium foil on one surface of cavity.
Surface resistances: when a solid and fluid are in contact there is surface resistance to the flow of heat, owing to a static layer of air next to the solid. The thickness of this layer depends on degree of air movement.
∴ external surface resistance $<$ internal surface resistance

Internal surface resistance (R_{si}) depends on surface emissivity and direction of heat flow.
External surface resistance (R_{so}) depends on degree of exposure (sheltered, normal, severe), surface orientation and type of construction.

Total resistance (R_t) of a compound structure:

$$R_t = R_1 + R_2 + R_3 + \ldots$$

Thermal transmittance, U (W/(m^2 K)): reciprocal of thermal resistance.

$$U = 1/R_t$$

Fabric heat loss, Q_c (W):

$$Q_c = UA(t_i - t_o)$$

where U = transmission coefficient (U-value) of the structure (W/(m^2 K))
A = area of the structure (m^2)
t_i = temperature of internal air (°C)
t_o = temperature of external air (°C)

Pattern staining occurs on the inside surface if the U-value is not uniform over the whole ceiling or wall.

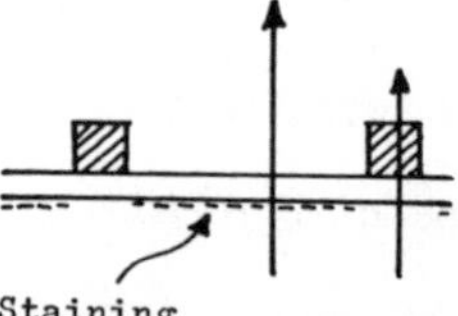

Thermal bridges permit the conduction of heat through part of an otherwise well insulated construction.

Heat losses through pitched roofs

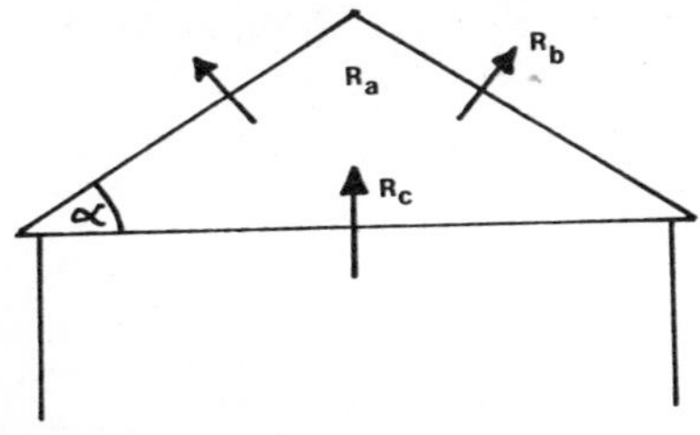

$$U = \frac{1}{R_c + R_a + R_b \cos\alpha}$$

where R_a = thermal resistance of loft space
$R_b = R_1 + R_{so}$ = thermal resistance of sloping portion
$R_c = R_2 + R_{si}$ = thermal resistance of ceiling
R_1 = thermal resistance of materials of sloping roof
R_2 = thermal resistance of materials of ceiling.

Temperature gradients across materials and structures

$$\Delta T = \frac{T \cdot \Delta R}{R_t}$$

where ΔT is the temperature drop across elemental resistance ΔR
T is the total temperature drop across the total thermal resistance R_t.

Homogeneous material

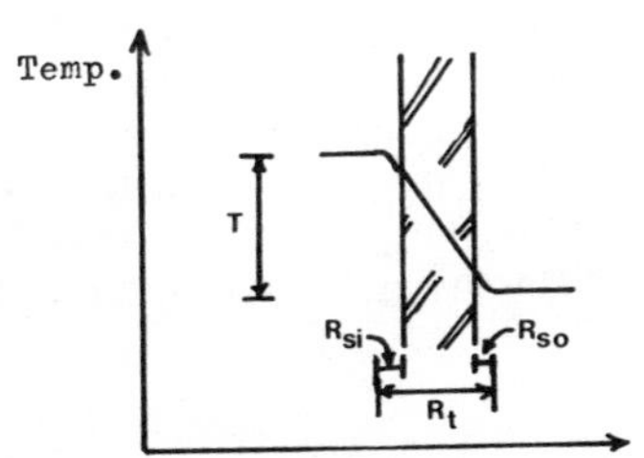

Compound structure

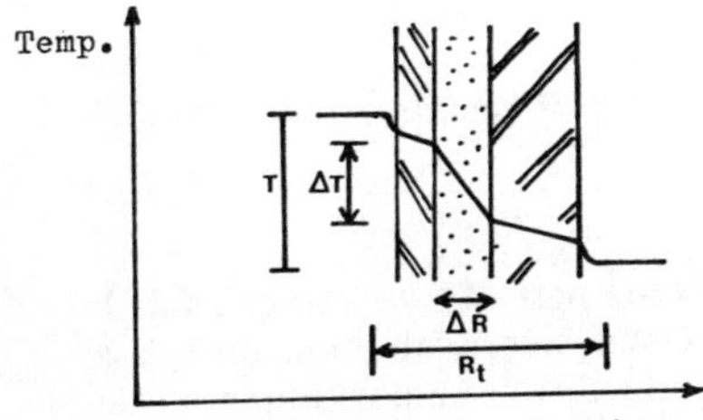

Example:

To study the temperature gradient in a wall construction:

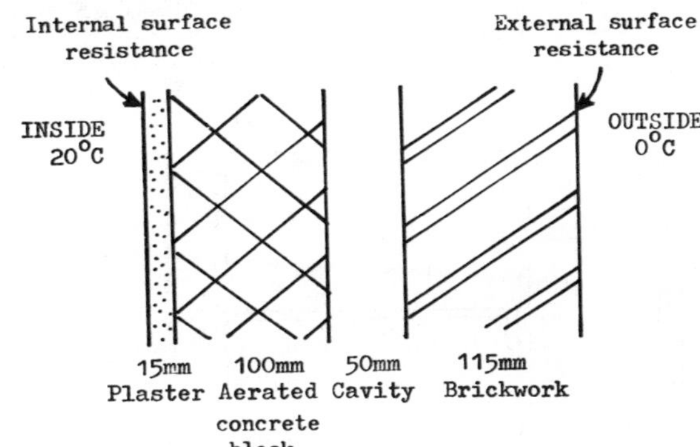

Section	Thickness m	Thermal conductivity, k W/(m K)	Thermal resistance, R m² K/W
Internal surface	–	–	0.123
Plaster	0.015	0.160	0.094
Concrete block	0.100	0.205	0.488
Cavity	–	–	0.180
Brick	0.115	1.180	0.097
External surface	–	–	0.053
			R_t = 1.035

∴ U-value = $1/R_t = 0.966$ W/(m² K)

Temperatures at section interfaces

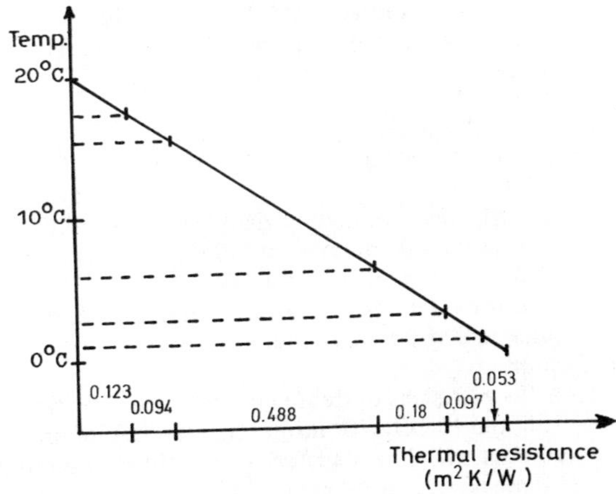

Thermal insulation standards

Refer: Building Regulations 1976.

U-values (W/(m^2 K))

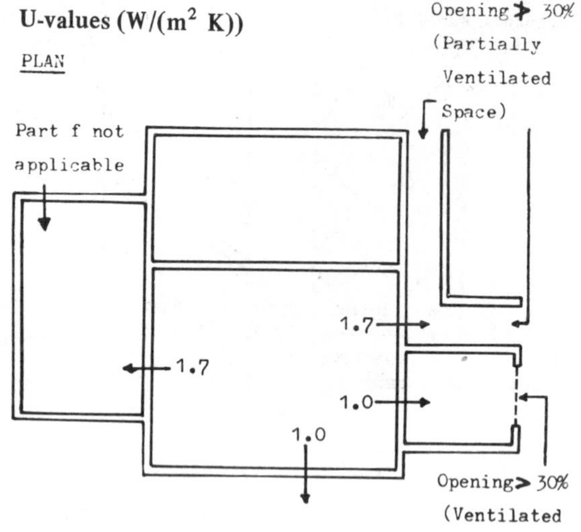

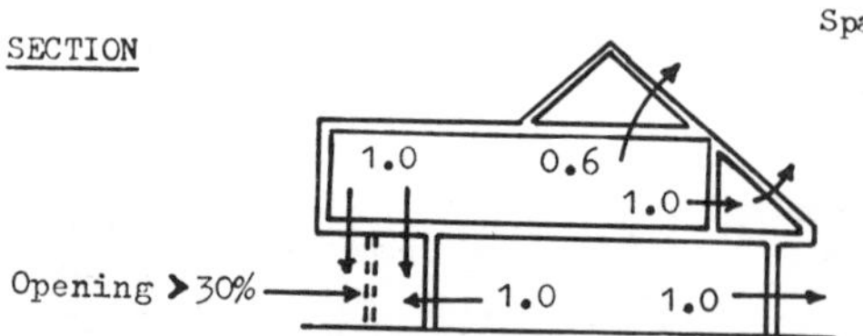

The calculated average U-value of perimeter walling (including any opening) shall not exceed 1.8 W/(m^2 K).

Example 'deemed-to-satisfy' provisions (Building Regulations 1976, Regulation F4, Schedule 11).

1. A cavity *wall* bonded by rigid leaves and filled with insulating material: minimum thickness of 37 mm mineral wool cavity fill.
2. A *roof* containing insulating material: minimum thickness of 60 mm mineral fibre (glass or rock) quilt.
3. A *floor* of slabs of dense concrete (not less than 150 mm thick) with insulating material in direct contact with lower surface of the floor: minimum thickness 21 mm mineral fibre mat.

CONDENSATION

Interstitial condensation occurs within the thickness of a building element.

Surface condensation occurs at the internal exposed surface of building elements.

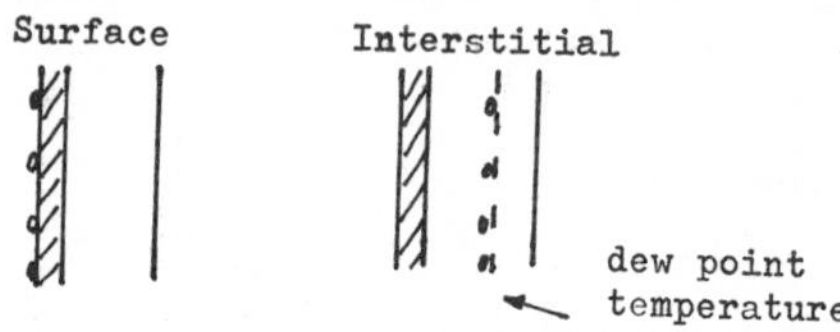

Dew point, t_{dp} (°C): the temperature at which moist air becomes saturated and condensation begins.

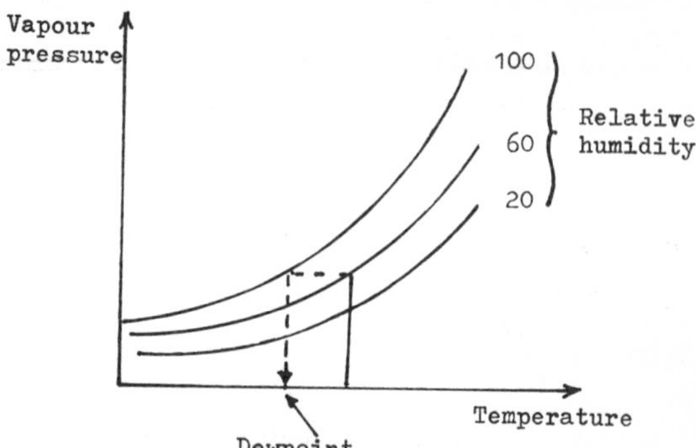

Relative humidity (r.h.) (%): the ratio of the actual vapour pressure to the saturation pressure at the same temperature, i.e. % r.h. at t°C. (Also known as *percentage of saturation vapour pressure.*)

Vapour diffusivity, d_v (g m/(MN s)): the mass of water vapour that passes through a unit cross-sectional area of material to a depth of 1 m per second under unit water vapour pressure gradient and at a given temperature.

Vapour resistivity, r_v (MN s/(g m)): reciprocal of vapour diffusivity.

$$r_v = 1/d_v$$

Vapour resistance, R_v (MN s/g): product of vapour resistivity and thickness.

$$R_v = r_v L = L/d_v$$

where L = thickness in metres.

Total vapour resistance (R_{vt}) of a compound structure:

$$R_{vt} = R_{v1} + R_{v2} + R_{v3} \dots$$

Vapour barriers are part of a constructional element through which water vapour cannot pass. The barrier should always be positioned on the warm side of the construction.

Example:

To study if condensation occurs in a washroom wall:

	Internal	*External*
Temperature	12°C	0°C
Relative humidity	77%	90%
Vapour pressure (from psychrometric charts)	1070 Pa	530 Pa

Internal/external temperature drop = T = 12 deg C.
Internal/external vapour pressure drop = 1070 − 530 = 540 Pa.

Conclusion. When the structural temperature falls below the dew point temperature (see table and diagram) some intermittent condensation would be expected.

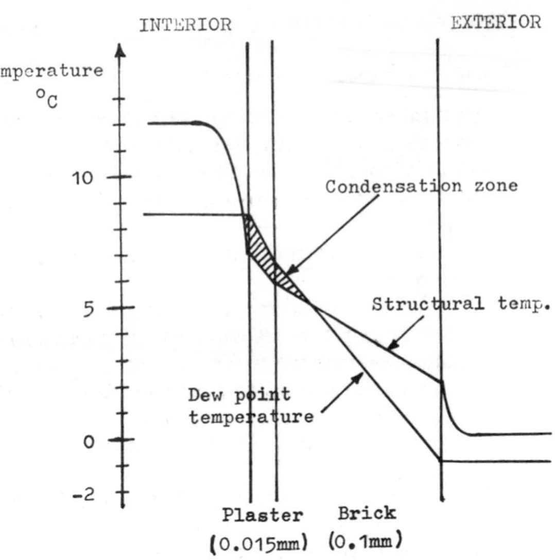

Section	*Thickness,* L m	*Thermal conductivity,* k W/(m K)	*Thermal resistance,* $\Delta R = L/k$ m² K/W	*Temp. drop,* $\Delta T = T \cdot \Delta R / R_t$ deg C	*Vapour resistivity,* r_v MN s/(g m)	*Vapour resistance,* $\Delta R_v = L r_v$ MN s/g	*Vapour pressure drop,* $\Delta P = P \cdot \Delta R_v / R_v$ Pa	**Dew-point temp. drop,* t_{dp} deg C
Int. surface			0.12	$12 \frac{0.12}{0.29}$ = 4.97	Negligible	Negligible	Negligible	Negligible
Plaster	0.015	0.5	0.03	$12 \frac{0.03}{0.29}$ = 1.24	60	0.9	$540 \frac{0.9}{4.9}$ = 99.2	8.4 to 6.8
Brick	0.100	1.1	0.09	$12 \frac{0.09}{0.29}$ = 3.72	40	4.0	$540 \frac{4.0}{4.9}$ = 440.8	6.8 to −1.0
Ext. surface			0.05	$12 \frac{0.05}{0.29}$ = 2.07	Negligible	Negligible	Negligible	Negligible
			$R_t = 0.29$	$T = 12.0$		$R_v = 4.9$	$P = 540$	

*from psychrometric charts

THERMAL COMFORT CRITERIA

Physical variables

1. **Air temperature** (measured with thermometer, hot-wire anemometer, bimetal strip).
   (a) 18°C to 22°C, depending on sex and age.
   (b) Avoid excessive vertical temperature gradients.
2. **Radiant temperature** (measured with globe thermometer).
   (a) Average wall temperatures $\not>$ 3 deg. C below air temperature to avoid stuffiness sensation.
   (b) Mean radiant temperature $<$ air temperature.
3. **Fresh-air level and movement** (measured with anemometer, Kata thermometer).
   (a) Minimum supply rate 18 m^3/h per room occupant.
   (b) Variable air movement 0.1 m/s to 0.2 m/s (freshness).
4. **Air humidity** (measured with hygrometer, sling or fan or Assmann psychrometer).
   (a) Relative humidity 30% to 70%.
   (b) High humidity produces a feeling of oppressiveness at high air temperature and chill at low air temperature.
   (c) Low humidity causes a sensation of parchedness and allows build-up of static electricity.

Personal variables

1. **Metabolic rate** (W/m^2): rate of heat production varies with the activity of the person (e.g. seated at rest 60 W/m^2; heavy manual work 250 W/m^2).
2. **Clothing** (clo units): insulation value of one clo corresponds to a thermal resistance of 0.16 m^2 K/W (e.g. heavy three-piece suit, long underwear 1.5 clo).
3. **Age:** basal metabolic rate reduced with age.
4. **Sex:** women prefer slightly higher temperatures.

Thermal comfort indices

1. **Dry bulb temperature, t_a (°C)**
   Alcohol or mercury in glass thermometer.

2. **Wet bulb temperature, t_w (°C)**
   t_a thermometer with bulb covered in saturated muslin. Lower than dry bulb temperature; difference is proportional to rate of evaporation and hence atmospheric water vapour.

3. **Mean radiant temperature, t_r (°C)**

$$t_r = \frac{A_1 t_1 + A_2 t_2 + \dots}{A_1 + A_2 + \dots}$$

   where $t_1, t_2 \dots$ are the surface temperatures of the areas $A_1, A_2 \dots$ respectively.

4. **Globe temperature, t_g (°C)**
   150 mm hollow, black sphere containing t_a thermometer. Apply correction factors to determine t_r.

$$t_g = \frac{t_r + (2.35\, t_a \sqrt{v})}{1 + 2.35\sqrt{v}}$$

   where v is the air velocity (m s^{-1}).

5. **Effective temperature**
   Combines air temperature, air movement and humidity. *Corrected effective temperature* also considers radiation (not now used).

6. **Equivalent temperature, t_{eq} (°C)**
   Measured by Eupathescope. Combines air temperature, air movement and radiation.

$$t_{eq} = 0.522\, t_a + 0.478\, t_r - 0.21(37.8 - t_a)\sqrt{v}$$

7. **Resultant temperature, t_{res} (°C)**
   Exists in two forms:
   (a) dry form – combines air temperature, radiation effects and air movement;
   (b) wet form – also includes humidity.

$$\text{Dry resultant temperature} = \frac{t_r + 3.17\, t_a \sqrt{v}}{1 + 3.17\sqrt{v}}$$

   Measured by thermometer at the centre of blackened 100 mm globe.

8. **Environmental temperature t_{ei} (°C)**
   Combines effect of radiation and air temperature.

$$t_{ei} = \tfrac{2}{3} t_r + \tfrac{1}{3} t_a$$

Comfort zone for sedentary occupation

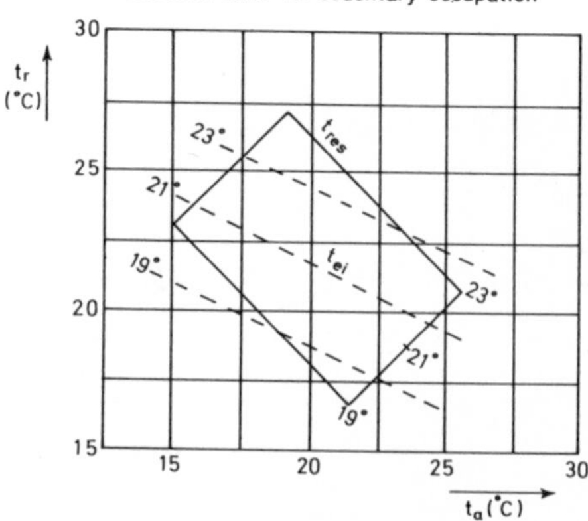

v = 0.1 m s^{-1} R.H. 40–70%

9. **Kata thermometer and index**
   Determines 'cooling power' of the environment. Now used as an omnidirectional anaemometer.

Acoustics: the science of sound.

NATURE OF SOUND

Sound is a sensation produced through the ear as a result of pressure variations in a medium, e.g. air, water.

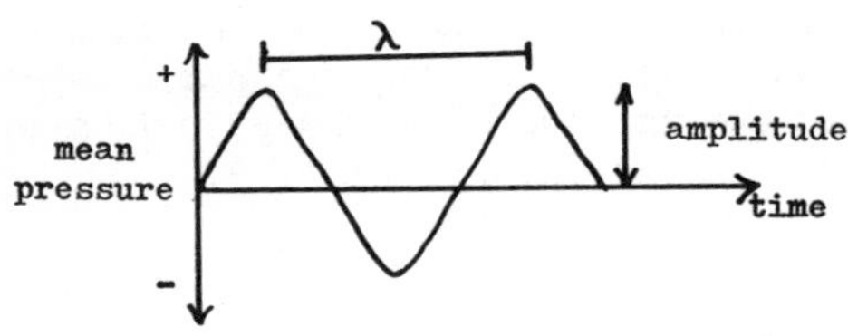

Velocity = wavelength × frequency $c = \lambda\nu$

Wavelength, λ (m): the distance travelled by the wavefront in one complete period of oscillation.

Velocity, c ($\mathrm{m\,s^{-1}}$): the velocity of propagation of a *longitudinal wave* through a material depends on:

1. Density (ρ) of the material
2. Temperature of the material
3. Modulus of elasticity (E) for *solids*, or bulk modulus (K) for *fluids*

For *solids*, $c = \sqrt{(E/\rho)}$ For *fluids*, $c = \sqrt{(K/\rho)} = \sqrt{(\gamma p/\rho)}$

where p is the pressure
γ is the ratio of the specific heat capacities.
Velocity of sound in air at one atmosphere pressure at 0°C is 332 $\mathrm{m\,s^{-1}}$.

Frequency, ν (hertz, Hz): the rate at which vibrations occur, i.e. the number of cycles per second.

Infrasonics (below 20 Hz)	0–1 Hz Wind effects on buildings 7 Hz Organs of human body resonate
Audible range (20 Hz to 20 kHz)	50 Hz Mains hum 164 Hz 'Big Ben' 440 Hz Pitch standard 1000 Hz GMT time signal
Ultrasonics (above 20 kHz)	Used in non-destructive testing. Produces chemical and heating effects.

Sound spectrum of a pure (or discrete) tone

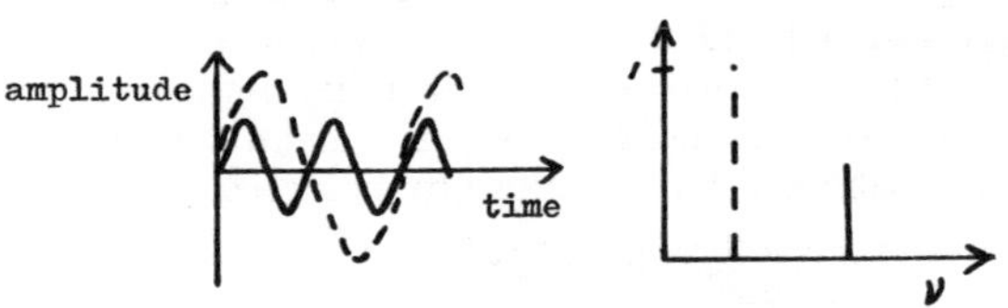

Sound spectrum of a periodic function

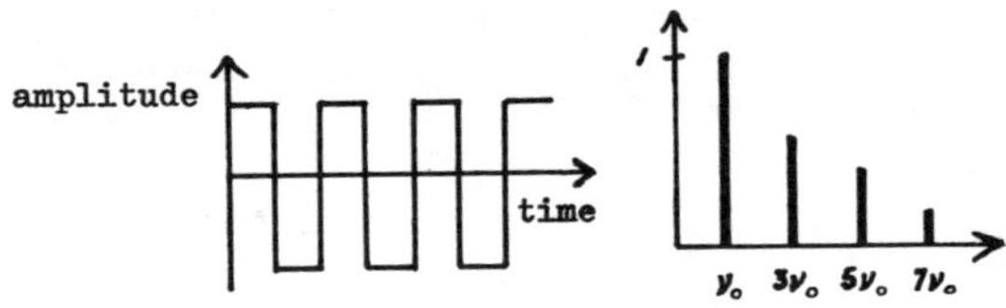

Sound spectrum of a broadband sound

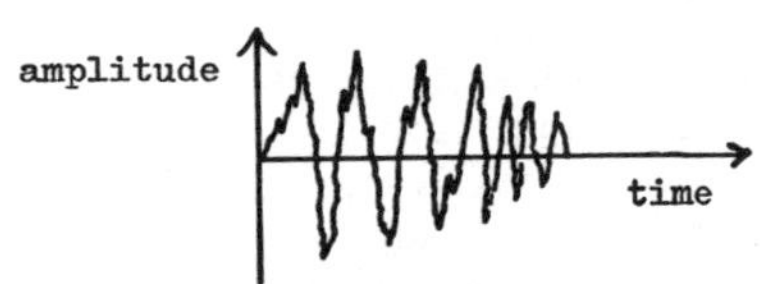

ANALYSIS OF SOUND

Sound power, P (W): the sound energy transferred per unit time. The power of an average voice in normal conversation is about 2×10^{-5} W.

Sound intensity, I ($\mathrm{W/m^2}$): the sound power radiated per unit area. If the sound is radiating equally in all directions the intensity (I) at a distance r from the source is:

$$I = \frac{P}{4\pi r^2}$$

Sound pressure, p (micro pascal, μPa): the slight variation of air pressure, above and below normal atmospheric pressure, caused by the sound. To avoid considering negative air pressures, the root mean square (r.m.s.) value is used:

r.m.s. value = 0.707 max. value

Note: crest factors for non-periodic functions.
The smallest r.m.s. pressure change, at a frequency of 1000 Hz, that the average young adult can detect is 20 μPa.
Sound intensities are proportional to the square of sound pressures, hence intensities do not involve r.m.s. values.

$$I = p^2/c\rho \quad \text{or} \quad I \propto p^2$$

where ρ = density of air
c = speed of sound in air.
$c\rho$ = characteristic impedance of air.

32 ACOUSTICS (2)

Decibel scale (dB)

A logarithmic ratio scale of two quantities, which corresponds closely to the way humans discern sound. When one quantity is standardised, the word 'level' is added to the scale, and denoted by the symbol L with an appropriate suffix.

Sound power level, L_W (dB):

$$L_W = 10 \log (P_1/P_0)$$

where P_1 is the sound power of the source (W)
P_0 is the reference power (10^{-12} W).

Sound intensity level, L_I (dB):

$$L_I = 10 \log (I_1/I_0)$$

where I_1 is the intensity of the sound (W/m^2)
I_0 is the reference intensity (10^{-12} W/m^2).

Sound pressure level, L_p (dB):

$$L_p = 20 \log (p_1/p_0) = 10 \log (p_1^2/p_0^2)$$

where p_1 is the measured r.m.s. pressure (μPa)
p_0 is the reference pressure (20 μPa), the threshold of hearing at 1000 Hz.

Threshold of feeling: $L_p = 120$ dB
Threshold of pain: $L_p = 140$ dB (at all frequencies)

Note: $L_p \approx L_I$

Addition of sound levels L_1 and L_2

$L_1 - L_2$	ΔL to higher level
0 to 1	3 dB
2 to 3	2
4 to 9	1
Above 10	0

Subtraction of sound levels, total sound level L_3; background level L_4

$L_3 - L_4$	ΔL from L_3
3	3 dB
4 to 5	2
6 to 9	1
Above 10	0

Sound-level meter consists of microphone, amplifier, r.m.s. rectifier, filters and meter display. Calibrate instrument before taking each set of readings, e.g. pistonphone.

Frequency analysis

Bandwidths: range of frequencies being analysed at any instant of time.

Octaves: bandwidths in which the higher frequency is double the lower limiting frequency (ν_1).

Standardised octave-band centre frequencies: 31.5, 63, 125, 250, 500 Hz; 1, 2, 4, 8, 16 kHz. Centre frequencies (ν_c) are geometric means, hence:

$$\nu_c = \sqrt{2}\nu_1$$

Octave-band analysis normally adequate for continuous sounds that do not have strong line spectra as a function of frequency.

Third octaves: narrower frequency bands used if greater detail is required. Results in a lower measured level due to less sound energy in the band. Correction added to narrower band level = 10 log (ratio of bandwidths).
Example: $10 \log (3/1) = 10 \times 0.477 = 4.8$ dB to give octave bands.

Frequency-weighting filters for sound-level meters

db(A) correlates fairly well with subjective assessment of loudness and is almost exclusively used.
dB(B) corresponds to the 70 dB equiloudness curve for pure tones.
dB(C) has a flat response for most of audio-frequency.
dB(D) used for aircraft-noise measurements.

ACOUSTIC CRITERIA

Loudness

A subjective quality that depends on both *frequency* and *intensity*.

Loudness level, L_N (phons): the loudness level of any sound, in phons, is numerically equal to the sound pressure level, in decibels, of a 1000 Hz pure tone, which an average listener judges to be as loud as the sound to be evaluated.

Equi-loudness contours:

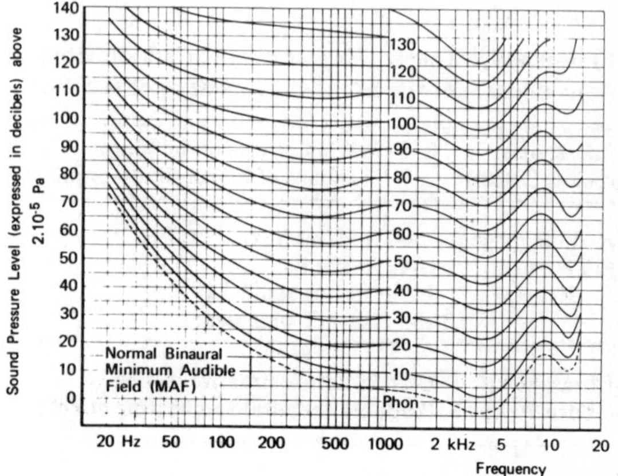

Loudness, N (sones) is the normal observer's auditory estimate of the ratio between the strength of the sound considered and that of a reference sound having a loudness level of 40 phons. Loudness units are directly proportional to the magnitude of the sensation experienced.
Loudness, in sones, is doubled each time the *loudness level* is increased by 10 phons.

$$N = 2^{(L_N - 40)/10}$$

Loudness index is a number determined by the geometric mean frequency and the pressure level of the octave band for a complex sound.

$$N_t = N_m + 0.3\ (\Sigma N - N_m)$$

where N_t = total loudness (sones)
N_m = greatest of the loudness indices
ΣN = sum of loudness indices of all the octave bands

Noise

Unwanted sound. Noise data should be corrected for relative humidity, temperature, reflected sounds and background noise, directional and frequency characteristics of the sound.

Perceived noise level, L_{PN} (PNdB): index of aircraft noise. The 'equinoisiness' curves replace the 'equiloudness' curves for aircraft noise.
'Loudness' (sones) results are replaced by 'noisiness' (noys).

Noise and number index (NNI): index of air traffic noise. Includes number of aircraft heard in a given period and average peak perceived noise level.

Traffic noise, L_{10} (18 hour): index of road traffic noise. The arithmetic average hourly value of noise level, dB(A), exceeded for 10% of the time, between the hours 06.00 and 24.00 on any normal weekday. Provides meaningful correlation with social response.

Corrected noise level (CNL): index of industrial noise. The noise level emitted from industrial premises corrected for tonal and impulsive character, intermittency and duration, in accordance with BS 4142.

Speech interference level (SIL): index of speech intelligibility with masking noise. The arithmetic mean of sound intensity of the interfering noise (dB) in the three octave bands between 600 and 4800 Hz.

Loudness index graph (with sone/phon conversion scale)

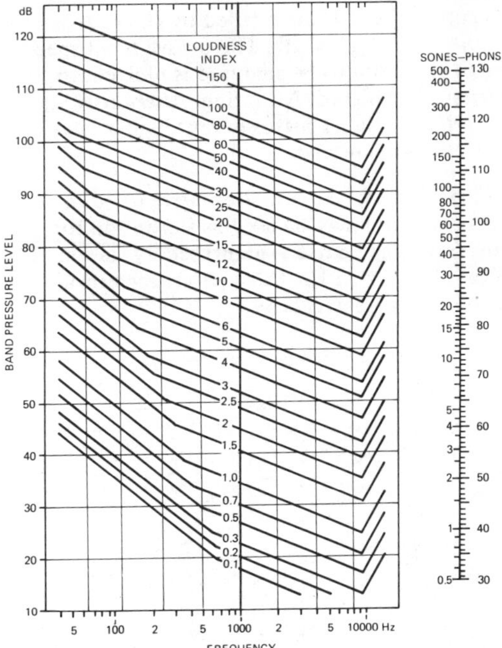

Equinoisiness curves

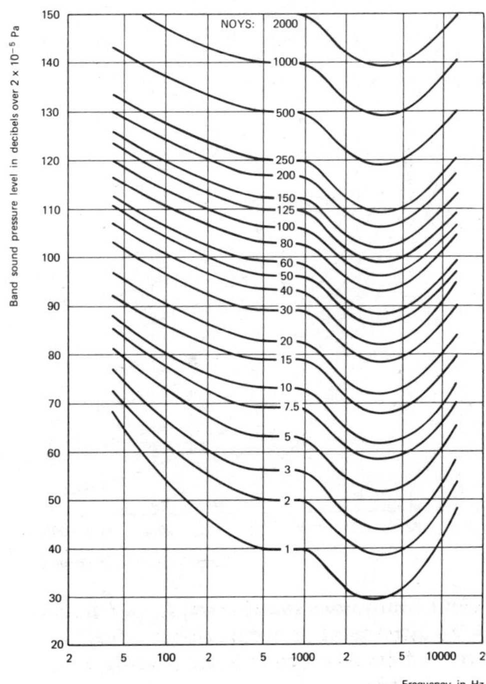

Noise criteria (NC) curves: index of maximum permissible background office noise, in octave bands, for different activities. The numerical rating of each curve is the SIL, identified by the sound pressure level at 1700 Hz. The octave band spectrum of the noise to be analysed is plotted, to determine the highest NC rating. *Alternative noise criteria* (NCA) curves permit higher low-frequency noise levels.

Noise rating (NR) curves: alternative to NC and NCA curves and may be used as a criterion for annoyance. Each curve is identified by its sound pressure level at 1 kHz. Corrections for various environments are necessary.

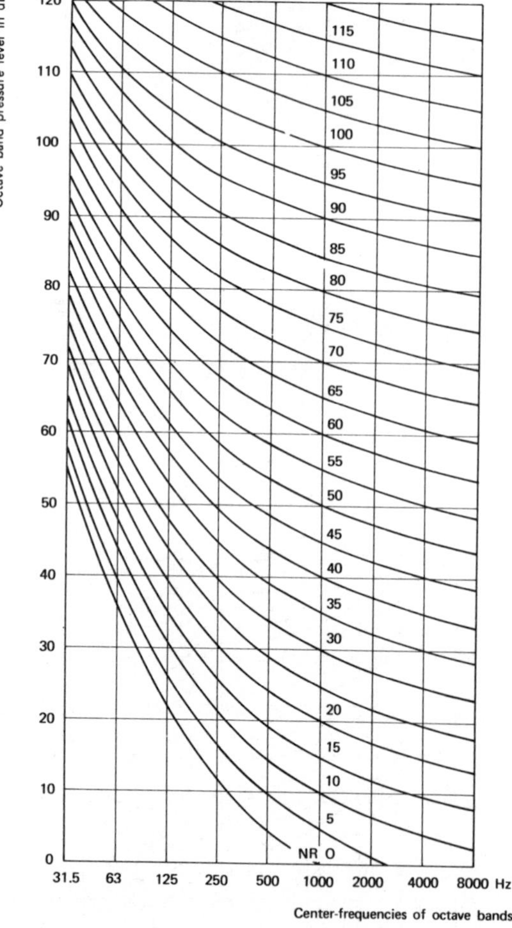

Equivalent continuous sound level, L_{eq} (T hours), **dB(A)**: an assessment of occupational noise exposure and duration, to indicate hearing damage potential.

$$L_{eq} = 10 \log \frac{1}{T} \int_0^T \left(\frac{p_A(t)}{p_o}\right)^2 dt$$

where L_{eq} is the equivalent continuous sound level (dB(A)) measured over time period T

$p_A(t)$ is the instantaneous A-weighted sound pressure (Pa) varying with time t

p_o is the reference sound pressure (Pa).

STEPWISE VARIATIONS

dB(A) — L_{eq}

RANDOM VARIATIONS

dB(A) — L_{eq}

The actual noise level in a given time period T is 'normalised' to a standard 8 hour *noise dose*. Doubling the time exposure reduces the L_{eq} by 3 dB(A) on the equal energy principle.

BUILDING ACOUSTICS

1. Provision of best conditions for listening—consider echoes, dead spots, reverberation time.
2. Elimination of noise from outside and inside building—consider sound *absorption* and *insulation*.

Reflection of sound

Sound may be reflected at both plane and curved surfaces, similar to light; but sound is *diffracted* around objects to a greater extent than light.

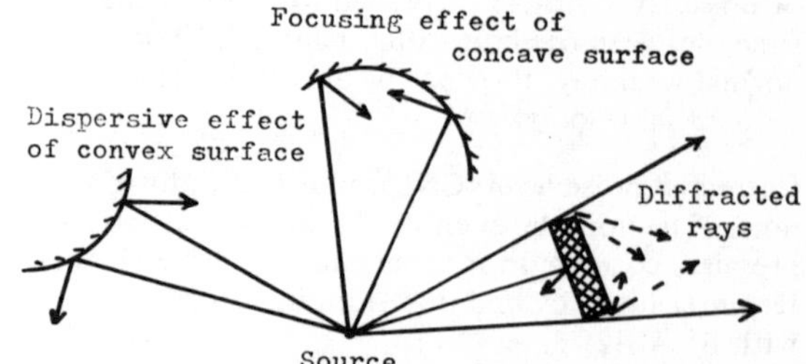

Echoes are sounds that have been reflected and arrive with such a magnitude and time interval after the direct sound as to be distinguishable as a repetition of it. Usually occurs when large concave surfaces are present, e.g. domed ceilings.

Sound absorption

Absorption is prevention of reflection of sound, and is mainly a function of the surface properties of the material.

Absorption coefficient of a material, α

$$\alpha = \frac{\text{energy absorbed by surface}}{\text{energy incident on surface}}$$

For a perfect absorber, e.g. an open window, $\alpha = 1.0$.

α is frequency-dependent, and may be influenced by the angle of incident sound.

Value determined using an impedance tube.

Types of absorbents

1. Porous materials
2. Membrane absorbers
3. Resonators

Absorption of sound in an enclosure

Reverberation time, T (seconds): time required for the average sound-energy density in an enclosure to decrease to 10^{-6} of the initial value (60 dB) after source has stopped.

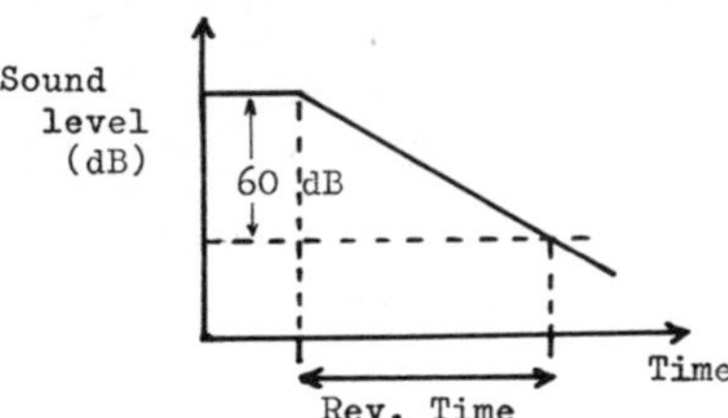

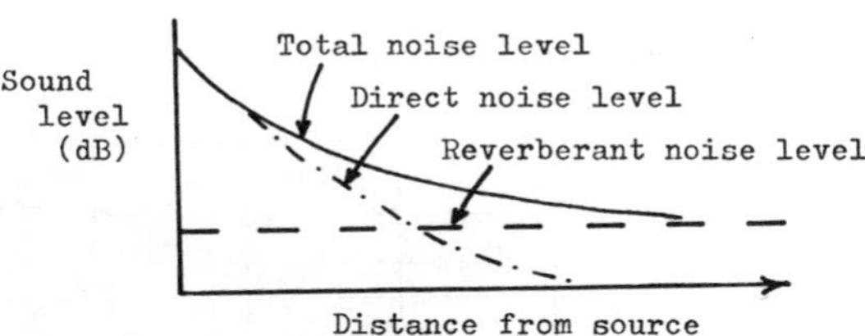

Subjectively, the reverberant energy from one transient sound should not mask the appreciation of successive transients. Too short a reverberation time 'deadens' room.

Sabine's Formula: $T = 0.16V/A$

where T = reverberation time (seconds)
V = volume of hall (m^3)
A = absorption units (m^2)
$= \alpha_1 s_1 + \alpha_2 s_2 + \alpha_3 s_3 + \ldots$
where $s_1 \ldots$ are the areas (m^2)
$\alpha_1 \ldots$ are the absorption coefficients

Sound insulation

Insulation is the prevention of sound transmission into an adjoining air space, and is mainly a function of the type of construction and surrounding structure.

Exclusion of outdoor noises

1. Zoning and siting: by effective town planning
2. Screening: by using sound barriers
3. Design: by raising insulation values, judicious location of noisy machines and isolation of vibrating plant.

Reduction of indoor noises

1. Airborne sounds, e.g. human voices, radio, etc.
2. Impact sounds, e.g. footsteps, noise from lifts, etc.

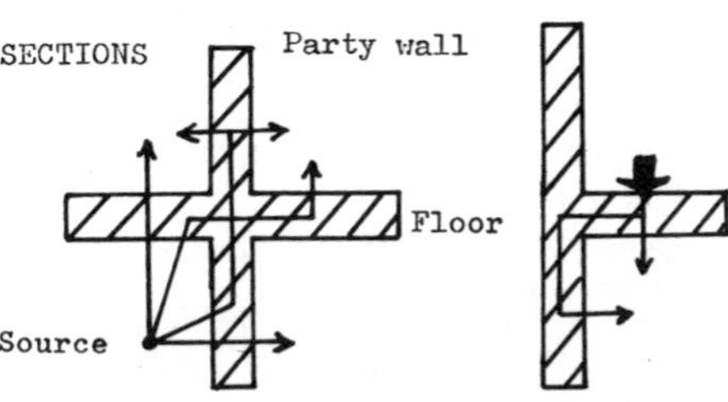

Airborne sound insulation

Factors influencing amount of insulation:

1. *Distance from source.* For each doubling of distance, attenuation is 6 dB for point source, 3 dB for line source.
2. *Mass effect.* Doubling mass of structure produces 5 dB reduction. But increasing mass to improve insulation rapidly becomes uneconomic.
3. *Uniform resistance to noise.* Insulation is only as good as the weakest point.
4. *Airtightness.* Small air-holes through partitions, ventilation ducts, and badly designed suspended ceilings can considerably reduce insulation.

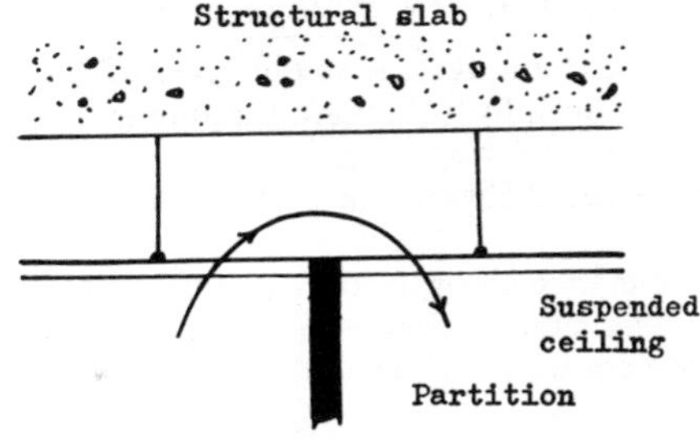

Approximate average sound insulation of windows
Wide open window: 5 dB.
Slightly open single window: 10–15 dB.
Closed openable window: 20 dB.
Sealed double window (6 mm, min. 230 mm air space, 6 mm): 40 dB.

Average sound pressure level, L, in a room is:

$$L = 10 \log \frac{p_1^2 + p_2^2 + \ldots p_n^2}{n p_o^2} \text{ dB}$$

where $p_1, p_2 \ldots p_n$ = r.m.s. sound pressures at n different positions in the room
p_o = reference sound pressure level.

Average sound pressure level difference, D:

$$D = L_1 - L_2$$

where L_1 = average sound pressure level in the source room
L_2 = average sound pressure level in the receiving room.

Sound reduction index, R: laboratory measurements (indirect transmission negligible).

$$R = L_1 - L_2 + 10 \log(s/A)$$

where s = area of test specimen (m^2)
A = total absorption in receiving room (m^2).

Normalised sound pressure level difference, D_n: field measurements.

$$D_n = L_1 - L_2 + 10 \log(T/0.5)$$

where T = reverberation time of receiving room (seconds) at 500 Hz.
0.5 = reference reverberation time.

Impact sound insulation

Normalised impact sound level, L_n:

$$L_n = L + 10 \log(A/A_o) = L - 10 \log(T/0.5)$$

where L = logarithmic average sound pressure level produced by a standard tapping machine in the receiving room
A = measured absorption in the receiving room
A_o = reference absorption (= 10 m^2)
T = reverberation time of receiving room.

Note. A correction of 5 dB should be added for third-octave results.

Insulation standards for dwellings

Building Regulations (English), 1976, Section G, specify limiting normalised levels of airborne and impact sound insulation for party walls of houses (*House Grade*) and walls and floors of flats (*Grade I and II*). Impact sound standards do not apply to walls.

The aggregate deviations of failures should not exceed 23 dB over the 16 specified frequencies. Divergencies in the unfavourable direction for *airborne* sounds are *below* the curve and for *impact* sounds *above* the curve.

Frequency range 100 to 3150 Hz. Below 100 Hz standing-wave patterns cause large variations in SPL across room.

Impact sound insulation grades

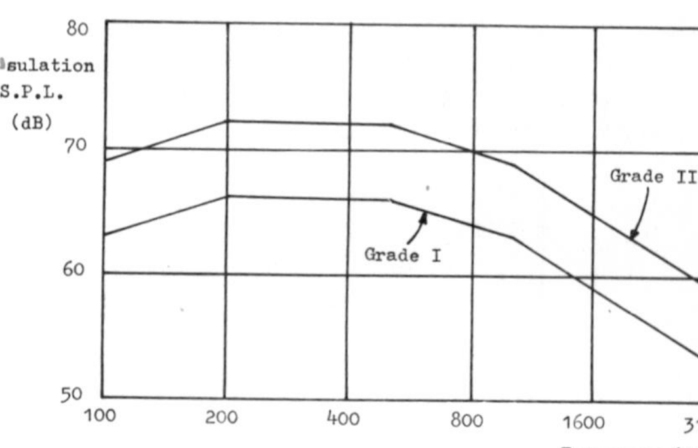

Airborne sound insulation grades

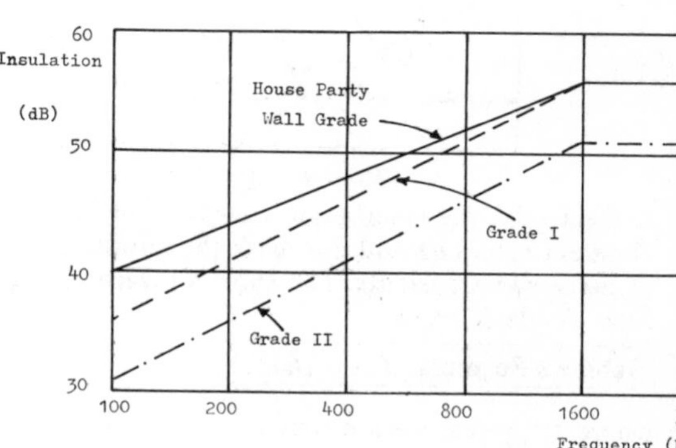

RAINFALL

Composition of water (H_2O)

Occurs in nature in various states of aggregation (water vapour, rain, fresh water, oceans and inland seas, ice caps and glaciers).

Properties of water

Maximum density: 1000 kg/m^3 occurs at 3.98°C

Specific gravity (sea water): 1.025 at 0°C

Specific heat capacity: 4187 J/(kg K)

Greater thermal capacity than any fluid except mercury

Latent heat of vaporisation: 2253 kJ/(kg K) at normal atmospheric pressure

Latent heat of fusion of ice: 334 kJ/(kg K) at normal atmospheric pressure

Surface tension: 73.5×10^{-3} N/m at 20°C

Boiling point: 100°C at s.t.p.

Freezing point: 0°C at s.t.p.

Potability: wholesome (spring, deep well, upland surface sources)
suspicious (river and stored rainwater)
dangerous (shallow well)

Water cycle

Evaporation, condensation and *precipitation.*
Rainwater absorbs carbon dioxide gas in the atmosphere to form carbonic acid (H_2CO_3).
Rainfall on limestone forms calcium bicarbonate (*alkaline*).
Rainfall on sandstone or peat forms *acidic* water.

pH value: measure of acidity. Defined as the logarithm of the reciprocal of the hydrogen ion concentration. Scale 0 to 14.
pH7 water is *neutral*; pH > 7 is *alkaline*; pH < 7 is *acidic*.

Hardness of water

A water is said to be hard when it is difficult to obtain a lather with soap.

Units of hardness: parts per million (p.p.m.) of calcium carbonate, irrespective of the actual salts present.
Less than 50 p.p.m. *soft water*; over 350 p.p.m. *very hard water.* 40% of the U.K. public water supply is between 200 and 300 p.p.m.

Types of hardness

1. *Temporary hardness*: can be removed by boiling. Due to calcium and magnesium bicarbonates. Forms scale and soft, non-adherent sludge in boilers.

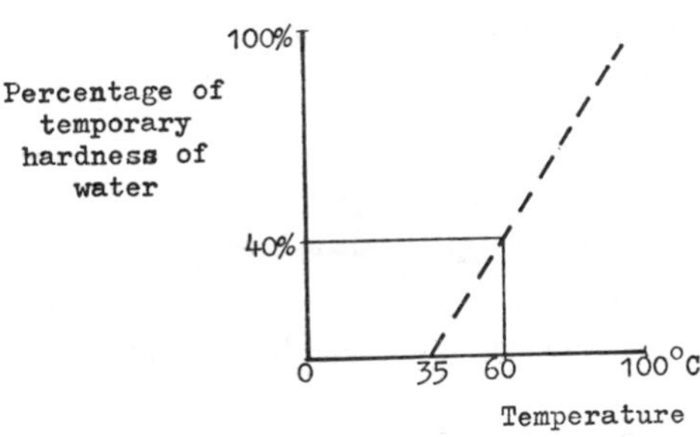

2. *Permanent hardness*: cannot be removed by boiling. Due to calcium and magnesium sulphates, chlorides and nitrates.
3. *Alkaline hardness*: due to calcium and magnesium carbonate or bicarbonate, or calcium hydroxide.

Softening of hard waters

Precipitation processes

Lime treatment
Removal of *temporary hardness* due to calcium bicarbonate:

Impurity	*Reagent*	*Precipitate*
$Ca(HCO_3)_2$	$+ Ca(OH)_2$	$= 2CaCO_3 + 2H_2O$
Calcium bicarbonate	Hydrated lime	Calcium carbonate

Removal of *temporary hardness* due to magnesium bicarbonate:

Impurity	*Reagent*	*Precipitate*	
$Mg(HCO_3)_2$	$+ 2Ca(OH)_2$	$= Mg(OH)_2$	$+ 2CaCO_3 + 2H_2O$
Magnesium bicarbonate	Hydrated lime	Magnesium hydroxide	Calcium carbonate

Note. Removal of magnesium bicarbonate requires twice as much lime as removal of calcium bicarbonate.

Removal of free *carbon dioxide* from the raw water:

Impurity	*Reagent*	*Precipitate*
CO_2	$+ Ca(OH)_2$	$= CaCO_3 + H_2O$
Dissolved carbon dioxide	Hydrated lime	Calcium carbonate

Removal of *permanent hardness* due to magnesium salts:

Impurity	*Reagent*	*Precipitate*	
$MgSO_4$		$CaSO_4$	
$MgCl_2$	$+ Ca(OH)_2 =$	$CaCl_2$	$+ Mg(OH)_2$
$Mg(NO_3)_2$		$Ca(NO_3)_2$	
Magnesium sulphate/ chloride/ nitrate	Hydrated lime	Calcium sulphate/ chloride/ nitrate	Magnesium hydroxide

The calcium salts formed by this process are removed by reaction with soda ash.

Soda ash treatment
Removal of *permanent hardness* due to calcium salts:

Impurity	*Reagent*	*By-product*	*Precipitate*
$CaSO_4$		Na_2SO_4	
$CaCl_2$	$+ Na_2CO_3 =$	$2NaCl$	$+ CaCO_3$
$Ca(NO_3)_2$		$2NaNO_3$	
Calcium sulphate/ chloride/ nitrate	Soda ash	Sodium sulphate/ chloride/ nitrate	Calcium carbonate

The by-products of this reaction are soluble, non-scale-forming sodium salts.
Precipitation methods are mainly used for large-scale treatment (except washing soda and bath salts), as exact proportions of chemicals must be added. On a domestic scale a glassy form of sodium metaphosphate may be suspended in a small cage in the cold water cistern. Sodium hexametaphosphate is mixed with water in washing machines.

Base-exchange processes

Water containing calcium and magnesium salts is passed through a bed of zeolites, essentially sodium aluminium silicates. The calcium and magnesium salts are replaced by the sodium to form soluble salts in the softened water. Zeolite can be regenerated by brine (NaCl). A zero hardness is attainable and independent of the quantity of reagent.

Distillation processes

Water is converted into steam which is allowed to condense. All impurities are removed and thus drinking water has to be artificially hardened. Expensive unless using solar energy.

Other methods

These include freezing salt water and electrical methods.

Waters for industrial boilers

Should not cause
corrosion (see section on 'Metals')
scaling (due to temporary hard waters)
priming (violent boiling resulting in projection of water droplets into steam pipes)
foaming (layer of bubbles forms on the water)
bumping (intermittent or explosive boiling; not very common)
or contain excess dissolved gases.

SOLAR DATA (Northern Hemisphere)

The apparent movement of the sun across the sky is due to the rotation of the earth about the sun and about its own axis.

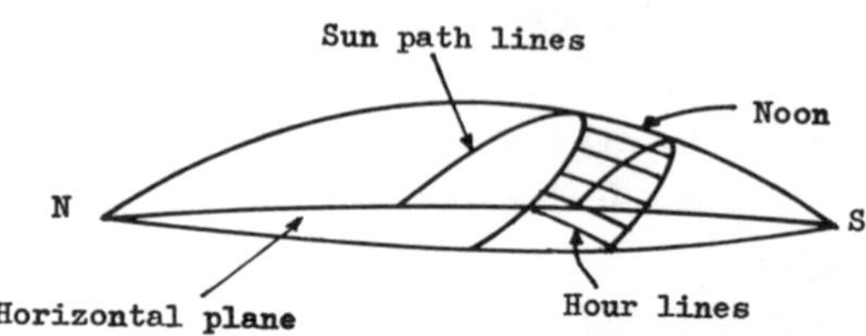

Highest elevation of the sun occurs at noon and is due south (max. 21 June, min. 21 December).
Sun can only pass through zenith position between Tropics of Cancer ($23\frac{1}{2}°$N) and Capricorn ($23\frac{1}{2}°$S).
In near-polar region, above Arctic Circle ($66\frac{1}{2}°$N) the sun never sets during some periods of the year.

Sun-path diagrams
1. Zenithal equidistant projection
2. Equatorial gnomic or perspective projections

Solar radiation on a surface depends on:
Latitude, orientation, vapour, time of day, season of year, atmospheric conditions.

Solar radiation in buildings causes:
Glare, deterioration of furniture and furnishings, increased heat loads, psychological effect of 'well-being'.

Solar attenuating devices:
1. External sunscreens
2. Venetian blinds (internal, external or between double glazing)
3. Glass (heat absorbing or reflecting)
4. Solar control film applied to windows
5. Heat reflecting curtains
6. Recessed windows
7. Light coloured external surfaces.

WIND

Wind effects on buildings

1. *External pressures.* Wind deflected around buildings is accelerated over the roof and sides, resulting in negative pressure (suction) on the surface.

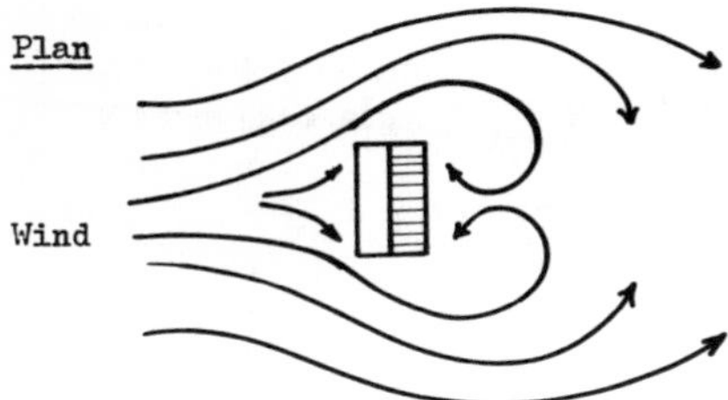

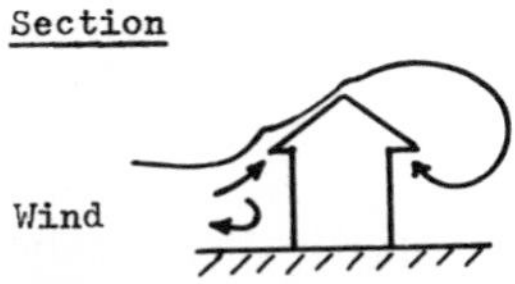

Gaps through or between buildings also cause increased air pressures.

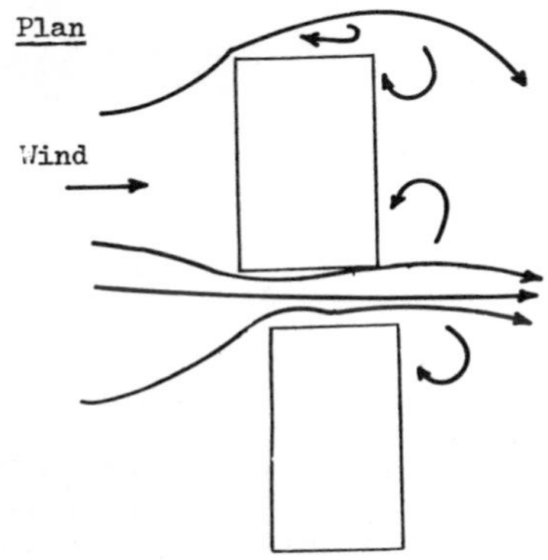

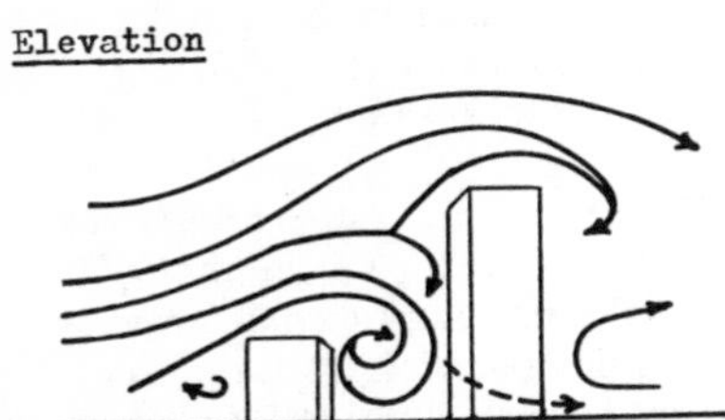

2. *Localised effects.* Regions around projections (e.g. parapets, chimney stacks, etc.) may be subjected to localised pressure variations.
3. *Internal pressures. Positive* internal pressure decreases the effect of external pressure and increases the effect of external suction. *Negative* internal pressure decreases the effect of external suction and increases the effect of external pressure.
   Internal pressures may result in damage to external cladding or internal partitions.
   Significant factors are:
   (a) permeability of walls or partitions
   (b) size and position of doorways and windows
   (c) presence of ventilators and chimneys.
   Note. Internal pressure and frictional drag may occur during the construction process.
4. *Building oscillations.* Relatively tall buildings may suffer from sway and wind-induced vibration which, although not structurally dangerous, may cause illness in the occupants of the building. These deflections can be reduced by structural damping or by architectural features, e.g. spiral strakes on large metal chimneys.

Wind velocity, V (m/s)
Classified on the Beaufort Scale:
Beaufort No. 0: wind speed 0–0.5 m/s, calm
Beaufort No. 10: wind speed 24.4–28.5 m/s, storm
Beaufort Nos 11 and 12 are rarely experienced.

Design wind speed, V_s (m/s)
Wind-speed contour maps of the U.K. are given in CP3, Chapter 5, and relate to a maximum 3-second gust. Design wind is adjusted for:
(a) surface roughness
(b) gust duration
(c) height of building

Air infiltration and ventilation

Infiltration: leakage of air through building due to structural imperfections.
Ventilation: designed supply of air into a building, e.g. naturally via windows, artificially via mechanical ventilation.

Natural forces causing air movement:
(a) wind pressure
(b) internal/external temperature gradient resulting in air density variations (Stack Effect).

Measurement of air infiltration: determined from the rate of disappearance of a tracer gas which has been completely mixed with air in the space.

$$C_t = C_o \exp(-Nt)$$

where C_t is concentration of tracer gas after time t ($m^3\ m^{-3}$)
C_o is initial concentration of tracer gas ($m^3\ m^{-3}$)
N is number of air changes in space during time t (s^{-1}).

Cements: hydraulic cements are used in concrete to provide a strong bonding material for the aggregate.
Types: Portland cements, high alumina cement, supersulphated cement.

PORTLAND CEMENTS

Manufacture

Raw materials
Calcium oxide, from chalk or limestone (CaO), C
Silica (SiO_2), S
Alumina (Al_2O_3), A
Iron oxide (Fe_2O_3), F

Process
Raw materials heated in a long, revolving, horizontal kiln to form *cement clinker*:
tricalcium silicate, C_3S
dicalcium silicate, C_2S
tricalcium aluminate, C_3A
tetracalcium alumino-ferrite, C_4AF
Clinker is ground with 4–7% gypsum ($CaSO_4 \cdot 2H_2O$) to prevent cement from setting too quickly on addition of water.

Setting and hardening

Hydration of cement produces a plastic cement-water paste, which expands on *setting* forming a three-dimensional crystalline framework and metastable gel. Initial and final setting times are determined using the *Vicat* apparatus. Drying and irreversible shrinkage occur on *hardening* when the cement stabilises forming a crystalline solid.

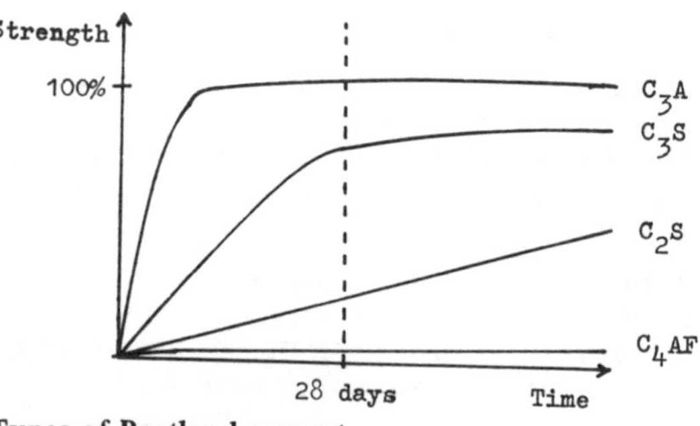

Types of Portland cement

1. *Ordinary Portland cement* (BS12)
   Suitable for most types of work.
2. *Rapid-hardening Portland Cement* (BS12)
   Develops high early strength.
   Permits early removal of formwork.
   Enables structure to be loaded earlier.
3. *Low-heat Portland cement* (BS1370)
   Heat of hydration produced at a lower rate.
   Lower proportion of C_3S and C_3A.
   Used in mass concrete work to avoid cracking.
4. *Sulphate-resisting Portland cement* (BS4027)
   Used in concrete exposed to *sulphate attack.*
   Very low aluminate content.
   Concrete should be dense with minimum water/cement ratio.
5. A number of other types also exist: extra-rapid-hardening PC, water-repellent PC, hydrophobic cement, masonry cement, white and coloured cements, Portland blastfurnace cement.

HIGH ALUMINA CEMENT

Manufacture

Melting bauxite ($Al_2O_3 \cdot 2H_2O$) and limestone in a reverberatory furnace and grinding the 'pigs' without gypsum into a fine powder.

Properties (BS 915)

Advantages
Gains strength quickly and has long setting period.
Predominant compound is monocalcium aluminate, CA.

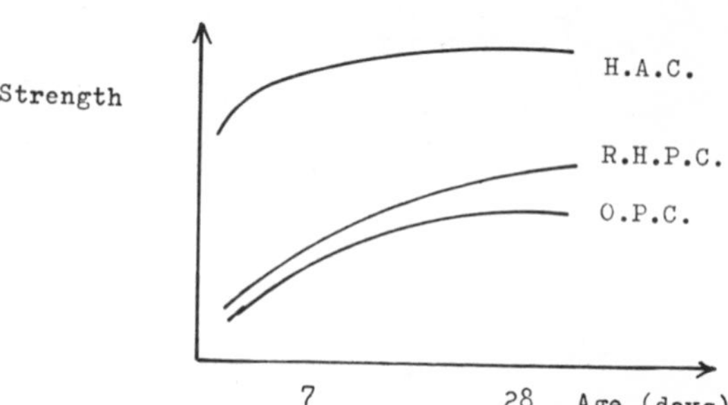

Resistance to acids is superior to that of OPC.
Is not affected by sulphates.
Does not disintegrate when subjected to heat.

Disadvantages
Great heat generated during setting and hardening.
HAC is more expensive than OPC because bauxite is imported from France.
'Flash' (instantaneous) set occurs if in contact with free lime.
Excessively high water/cement ratios nullify the advantages of this cement, particularly if hot and humid conditions persist during construction *or* at any time in service. Certain recent building failures have shown that HAC concrete can undergo *conversion* (a chemical change) accompanied by loss of strength and reduction of resistance to chemical attack. (Refer: BRE Current Paper No. 58/74.)

SUPERSULPHATED CEMENT

Manufacture

Grinding 90% blast furnace slag with 10% calcium sulphate and Portland cement clinker.

Properties (BS 4248)

High resistance to acids and alkali solutions.
Low early strength development, especially in cold weather.
Careful mixing and curing required.
Deteriorates rapidly if stored in damp conditions.

AGGREGATES

Include gravels, crushed stones and sands.

Types

1. *Natural*: relative density about 2.6, e.g. granite, flint, limestone etc.
   Used where strength, resistance to wear or impervious material is required.
2. *Artificial*: low relative density, e.g. clinker, foamed blast furnace slag, heated expanded clay etc.
   Used in lightweight concrete, provides good heat insulation.

Characteristics

1. *Durability*: hard, non-decomposable material which does not change volume on weathering or affect reinforcement.
2. *Clean*: free of organic impurities and dust.
3. *Shape*: rounded, irregular or angular.
4. *Texture*: glossy, smooth, granular, rough, crystalline, honeycombed.
5. *Size*: *fine* aggregates pass through 5 mm BS sieve, but *coarse* aggregates will be retained.
6. *Grading*: aggregates used for concrete should have a regular gradation of particle sizes to reduce resultant voids, increase workability and cement/aggregate ratio.
7. *Weigh batching*: more satisfactory than volume batching, owing to bulking effect.

Tests on aggregates

Obtain samples using a riffle box or quartering bar.
1. Field settling test for silt and clay.
2. pH test for acidity.
3. Moisture content: oven drying, siphon can, calcium carbide, electrical and dilution methods.

CONCRETE

Major constituents

Cement, aggregate and water: the cement and water form a paste, which coats the aggregate particles and fills the voids between them, resulting in a material that hardens with time.

Proportions of ingredients

Specification is by weight:
'1:2:4' means 1 of cement, 2 of fine aggregate, 4 of coarse aggregate.
'0.6 water/cement ratio, means 0.6 kg water to 1 kg cement.

Properties of concrete

Durability: to withstand severe weather conditions, chemical attack and abrasion.

Strength
(a) Tensile strength: low and of little use.
    Compressive strength: fairly high.
(b) Concrete gains in strength with time. Most mixes are designed to reach a minimum specified strength after 28 days, and the strength after 7 days should be approx. two-thirds this strength.
(c) The rate of gain of strength depends on the temperature at which it is *cured*.
(d) Depends on the type of aggregate used.
(e) Depends on air voids in the mix; 1% air voids can result in 5% decrease in ultimate strength.
(f) Low water/cement ratio required to produce dense, impervious material, but reduces workability.

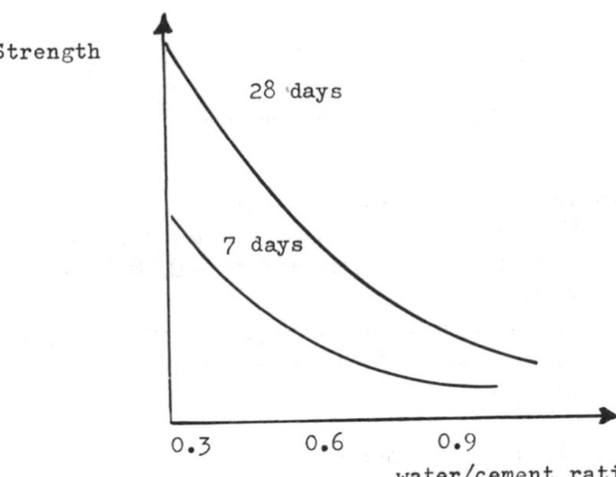

An indication of strength is given by crushing test cubes to destruction at 7 and 28 days.

Workability: the amount of work needed to fully compact the concrete. To improve workability:
(a) increase the water/cement ratio, but this reduces strength;
(b) use a well-graded aggregate which has the largest particle size and hence smaller surface area;
(c) use a smooth, rounded aggregate to reduce internal friction.

An indication of workability is given by the *slump test, compaction factor test* or the *'V-B' consistometer test.*

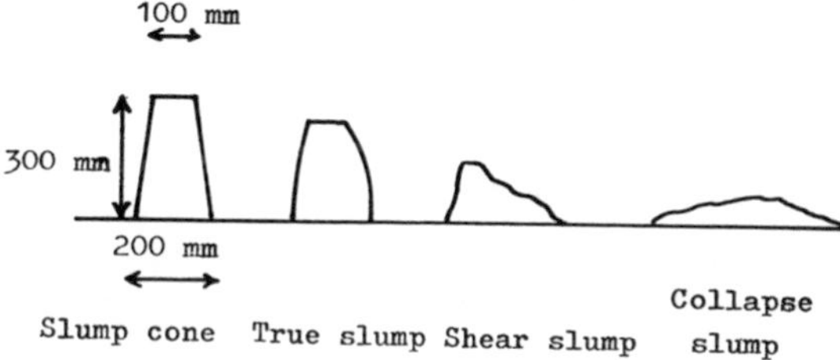

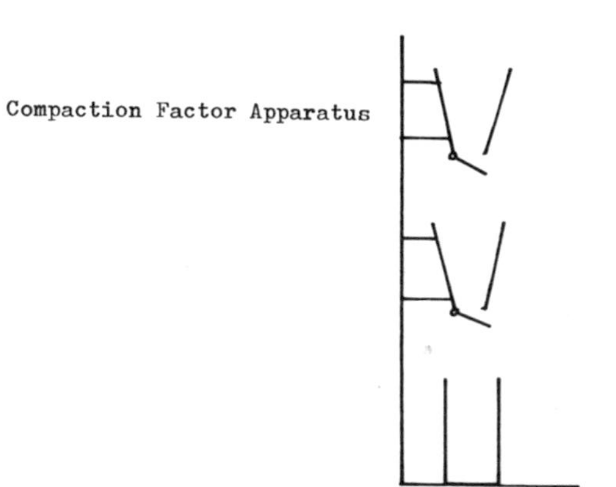

$$C.F. = \frac{\text{Wt. of partially compacted concrete}}{\text{Wt. of fully compacted concrete}}$$

Typical values of workability
(a) High-strength reinforced concrete sections compacted by vibration: C.F. = 0.78–0.85, slump = 0–25 mm.
(b) Normal reinforced concrete sections compacted by vibration: C.F. = 0.85–0.92, slump = 25–50 mm.

Admixtures for concrete

1. *Accelerators*: mainly used for cold-weather concreting; e.g. calcium chloride (should not be used in prestressed concrete).
2. *Air-entraining admixtures*: entrain air in concrete as a suitable bubble (about 1 mm) to improve workability of fresh concrete and frost resistance of hardened concrete; e.g. vinsol resin.
3. *Set retarders*: extend workability of concrete for long-distance pumping and for work in hot climates; e.g. lignosulphonates or hydroxycarboxylic acids.
4. *Plasticising admixtures*: improve concrete workability by reducing the surface tension of the water; e.g. Febmix.
5. *Special admixtures*: e.g. colouring, corrosion-inhibiting and fungicidal admixtures.

Note: Calcium chloride or admixtures containing calcium chloride must never be added to extra-rapid-hardening, sulphate-resisting, masonry or high-alumina cements.

LIGHTWEIGHT CONCRETES

Concrete with density between 200 and 1800 kg m^{-3}.

Types

1. *No-fines concrete*: concrete contains only coarse aggregate.
2. *Lightweight-aggregate concrete*: cellular or porous aggregate replaces coarse aggregate, e.g. furnace clinker, foamed slag, sintered pulverised fuel ash (p.f.a.), exfoliated vermiculite, expanded clay or perlite.
   Blocks can be solid, cellular or hollow, with ends plain, tongued-and-grooved, or double-grooved.
3. *Aerated concrete*: concrete containing bubbles of gas produced either by chemical means or by an air-entraining agent.

Advantages

1. Good thermal insulation and fire resistance.
2. Good sound absorption properties.
3. Reduction in dead loads of building.
4. Lower haulage and handling rates.
5. Faster building rates.
6. Provides a use for industrial wastes.

Disadvantages

1. Reduction in compressive strength and modulus of elasticity.
2. Susceptible to handling damage.
3. High moisture movement.

BUILDING LIMES

Limes mixed with cement give a *mortar* added workability. (*Note*. A *flash-set* occurs if added to high alumina cement.)

Manufacture

1. *Calcination*: chalk or limestone ($CaCO_3$) is burnt to quicklime (CaO).
2. *Slaking* of quicklime forms a hydrated lime:
   $CaO + H_2O \rightarrow Ca(OH)_2$
   Slaking, until recently, was done on site. Today, hydrated lime is delivered in powdered form. Incomplete slaking may cause *blowing, pitting* or *popping*.
3. Lime is marketed either in lump form, or as lime putty, or as hydrated lime powder.

Types

Quicklimes, hydrated limes, semi-hydraulic and eminently hydraulic (blue lias) limes.
Magnesium (non-hydraulic) limes contain up to 35% magnesium oxide.

Uses

Mortars, renderings, plasters, calcium silicate bricks.

PLASTERS

Manufacture

Crushed rock gypsum ($CaSO_4 \cdot 2H_2O$) is heated to 140°C to form a *hemihydrate*. Subsequent heating to 300°C results in *anhydrous calcium sulphate*.

Types (BS 1191)

1. *Class A Plaster*
   Plaster of Paris: hemihydrate.
   Rapid set; used for patching.
2. *Class B Plaster*
   Retarded hemihydrate gypsum plaster.
   Addition of a retarder in the form of a breakdown product of keratin to delay set.
   Suitable for general plastering.
3. *Class C Plaster*
   Anhydrous gypsum plaster.
   Slow setting without the addition of *accelerators*, e.g. zinc or potassium sulphate, alum: 0.5 to 1% by weight.
4. *Class D Keene's or Parian*
   Anhydrous gypsum plaster.
   Formed by *calcining* at 370°C and finely grinding.
   Used as a finishing plaster.
5. *Special Plasters*
   Acoustic plasters.
   Barium-based plasters for X-ray rooms.

PLASTERING MIXES

Combinations of Portland cement *or* gypsum plaster, lime, sand or lightweight aggregate, depending on the nature and hardness of the finished surface. Applied to a suitably prepared surface as two undercoats and a finishing coat. Each coat must be allowed to dry thoroughly to avoid excessive shrinkage.

Plastering defects

Refer: Department of the Environment Advisory Leaflet No. 2.

Blistering, bond failure, cracking, crazing, efflorescence, flaking and peeling of final coat, grinning and irregularity of surface texture, popping or blowing, recurrent surface dampness, rust staining, softness or chalkiness.

Plastering defects may be due to causes other than the use of faulty plastering materials or techniques, e.g. cracks in a plastered ceiling may be due to deflections of the construction.

PLASTERBOARD

Plasterboard consists of a core of set gypsum sandwiched between two stout lining papers. It can be fixed by nailing to wood studs or joists. Joints should be *scrimmed* before the application of 1 or 2 finishing coats of plaster. (Wood laths for plaster are no longer used.)

Types

Gypsum baseboard, wallboard, lath, plank.

CLAY PRODUCTS

Clay deposits are formed by the mechanical and chemical disintegration of surface rocks. The three main *clay minerals* are kaolinite, illite and montmorillonite.

Bricks

Clay types
Commons: suitable for general building purposes.
Facings: attractive appearance.
Engineering bricks: high strength and low water absorption.

Manufacture
After 'winning' the clay from quarries, holes or pits, it is crushed and ground. Water is added to produce the correct degree of plasticity. Materials to improve combustion during *firing* or appearance are also added.

Moulding to shape
1. Semi-dry process.
2. Wire-cut process.
3. Hand-made process for special shapes.

Firing. 4 stages occur in both the 'car-tunnel' and 'Hoffman' kilns:
1. Drying (150°C) to remove excess water.
2. Heating (900°C) to decompose carbonates and oxidise iron compounds.
3. Vitrification (900 to 1300°C) to cause localised sintering of clay and melt metallic oxide fluxes.
4. Cooling and the formation of a glassy crystalline coherent solid.

Size (metric)
It should be possible to hold a brick in one hand.
215 × 102.5 × 65 mm plus 10 mm joint.

Shape (BS 3921)
Solid with or without frogs; perforated; cellular; standard specials.

Properties
1. Wide variety of colours and textures.
2. Strength: 2.76–175 N/mm^2.
3. Water absorption may result in frost damage, efflorescence, crystallisation or sulphate attack.

Note. *Calcium silicate bricks* (sandlimes) are *not* clay-based, but are made by steam autoclaving, under pressure, a mixture of silica and lime. They have lower strength than clay bricks.

Clay pipes

Clay and 'grog' is mixed with water to form a sufficiently plastic mixture that can be extruded. The pipes are dried and fired in tunnel or intermittent kilns.

Types
1. *Porous*
   Used for land drainage.
   Laid unjointed.
2. *Vitrified*
   Vitrification gives the surface a glazed, impervious finish.
   Used for surface or foul water drains.
   Manufactured with socket and spigot (O-ring, mortar or tarred-gasket joints) or plain ends (sleeve joint).

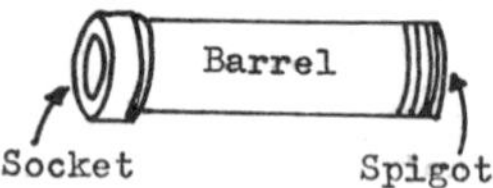

Clay roofing tiles

Hand or machine made in various sizes and shapes. Both nail holes and nibs may be present. Design should minimise capillary action between adjacent tiles. Good weather resistance and colour fastness.

Types
1. *Double-lap* (plain tiles)
   Overlap the two courses below and butt against adjacent side tiles.
2. *Single-lap* (interlocking tiles)
   Have a side-lap of about 25 mm.
   Edges ribbed or grooved to reduce rain penetration.
   Examples are Pantile or Roman varieties.

Clay blocks

Extruded hollow units made from diatomaceous earth. Easily cut and nailed.

Types
1. Loadbearing.
2. Non-loadbearing.

Ceramics

Terracotta
A fired sedimentary clay.
Very resistant to atmospheric conditions.

Faience
Glazed terracotta.
Used for tiles and plaques.

Earthenware
A blended clay and limestone.
Used for wall tiles (glazed type).

Fireclay
Strong with good fire resistance.
Used for refractory bricks.

Vitreous china
Very low water absorption.
Used for sanitary fittings.

Porcelain
Higher quality than vitreous china.
Used for electrical insulators.

WINDOW GLASSES

Glass: a vitreous silicate produced by heating (1500 °C) silica, sodium carbonate, alumina, limestone and other substances.

Manufacturing processes

1. The *float* process: optically flat glass formed on the surface of molten tin.
2. The *flat-drawn* process: glass may contain slight distortions.
3. The *rolled* process: for patterned and wired glass.

Modified glasses

1. *Additions* to the basic mix: ferrous oxide produces *heat absorbing* glass, which selectively absorbs near infra-red energy without greatly reducing the transmission of visible radiation, but gives glass a bluish-green coloration. Other chemicals form bronze and grey glasses.
2. *Coated glasses*: metallic or organic coatings produce a *heat reflecting* glass.
3. *Laminated glasses*: two sheets of glass cemented together with an organic interlayer form a *safety* (security) glass.
4. *Special glasses*: include toughened glass, patterned glasses for privacy, wired glass for fire resistance.

Properties

Light transmission
Low light transmission reduces task illuminance and lighting quality (including glare).

Thermal transmission

(a) *Heat losses from rooms*
Glass has a low coefficient of thermal conductivity (k = 1.05 W/(m K)). Doubling glass thickness only increases the overall thermal resistance by approximately 3%.
Double glazing with an air space reduces the thermal transmittance (U-value). Optimum separation of panes is 20 mm. Double glazing reduces condensation and down-draughts.
Types of double glazing include 'secondary window', coupled windows, sealed units.

(b) *Solar heat gains*
Can be predicted using graphical methods or equations.
May be reduced using solar-control glasses (heat absorbing or reflecting). *Note*: heat absorbing glasses can also reradiate heat *into* the building.

Acoustic insulation
Sound transmission through glass depends on the frequency of the sound. Resonances occur at low frequencies. Typical reduction for 6 mm thick single glazing in an openable window is 30 dB at 1000 Hz. Acoustic insulation can be improved by:
(a) increasing thickness of glass panel. Double thickness gives approximately 3 dB attenuation (little practical use).
(b) adopting a double-leaf construction. Sound insulation increases with width of air space (min. 100 mm) and improved properties of its perimeter surfaces.

Weather resistance
Static fatigue occurs in glass, i.e. it is weaker when subjected to long-duration loads than short-duration loads and thus should be designed to a lower breakage stress when snow-loaded. Design for wind-loading is based on a 3 second gust.
Under extreme climatic conditions the glass in sealed double-glazing units can fracture owing to variations in temperature and pressure of the filling air.
Glazing sealants, e.g. polysulphides and silicones, should eliminate rain penetration but permit movement in the joint.

Metals: ferrous and non-ferrous types.

FERROUS METALS

Most ferrous materials are *alloys*: a homogeneous multicomponent solid or liquid in which the primary component is a metal.
Occurs as *iron ore*: iron oxides or carbonates with small quantities of carbon, manganese, magnesia, silica, alumina, lime, phosphorus and sulphur.

Manufacture

Iron ore is smelted in a blast furnace to form *pig iron*, which is refined to produce cast irons, wrought iron, and steel.

Cast iron
Alloy of iron with > 1.7% carbon.
Hardest and most brittle iron.
Possesses low tensile strength (not used structurally).
Can be welded with oxy-acetylene.

Wrought iron
Alloy of iron with < 0.25% carbon.
Comparatively soft and malleable.
High resistance to corrosion and shock.
Similar to mild steel.

Steel
Pig iron with the impurities removed and with accurately controlled addition of other elements, e.g. chromium, manganese, nickel, tungsten and vanadium.

1. *Mild steels* contain < 0.5% carbon and are used in construction.
2. *Hard steels* contain between 0.5 and 1.7% carbon.
3. *Alloy steels*
   Include stainless steels.
   Long 'maintenance-free' life.
   Resistant to atmospheric corrosion.
   High elastic limit.
   18/8 (Cr/Ni) is suitable for internal use.
   18/10/3 (Cr/Ni/Mo) is suitable for external use.
   As carbon content increases, strength and hardness increase but ductility and ease of welding decrease.

Steel for reinforcement of concrete

Provides tensile strength that concrete lacks.
Withstands shear stresses in beams (greatest nearest support).
Reduces surface cracking in mass concrete structures.
Should be free of scale, loose rust, grease, oil or mud, which impairs bonding of steel to the concrete.
Bars may be round or deformed (e.g. ribbed).
Bars should be securely tied together with 16 s.w.g. soft iron wire.

Concrete lintel without reinforcement

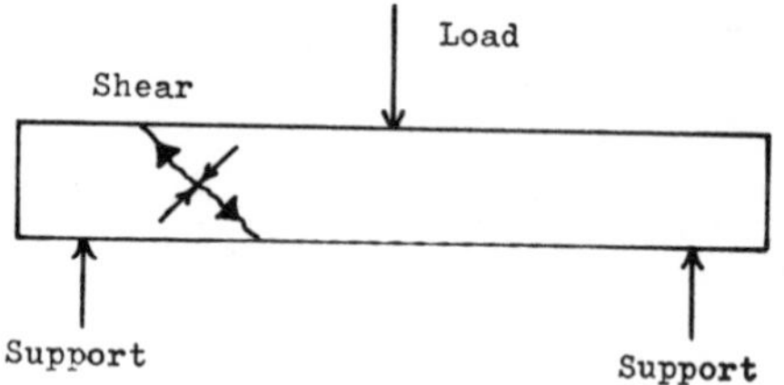

Reinforced concrete lintel

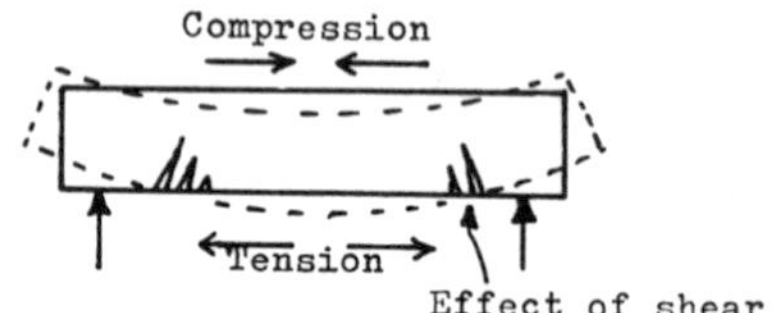

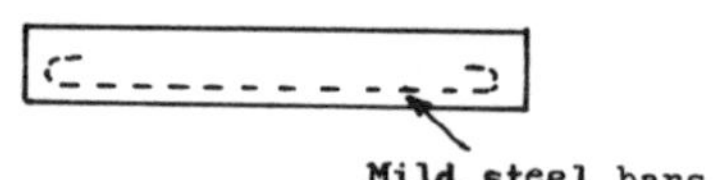

Steel (tendons) for prestressing of concrete

Contains higher percentages of carbon than reinforcing steels.
Work hardening the steel (by drawing it through a die) increases its tensile strength.

NON-FERROUS METALS

Copper (Cu)

Occurrence
In ores containing copper, iron and sulphur.

Properties
Very ductile and malleable.
Good electrical and thermal conductivities.
Low tensile strength.
Develops a protective *patina* on weathering.
Joined by welding, brazing and soldering.

Uses
Electrical cables, pipes, lightning conductors, d.p.c.s, roof coverings, wall-covering finishes, flashings.

Copper-base alloys

1. **Brasses:** alloys of copper and < 50% zinc (Zn).
2. **Bronzes:** alloys of copper and tin (Sn) and small quantities of other metals e.g. Ni, Pb, P, Zn. Harder and stronger than brasses, but less ductile.

Lead (Pb)

Occurrence
Main source is galena (lead sulphide).

Properties
The most dense common metal (S.G. = 11.34).
Low melting point (327 °C) and high thermal movement.
Low tensile strength, malleable and easily cut.
Forms a protective coating of lead carbonate on exposure to the atmosphere.
Attacked by weak acids.
Impervious to moisture but liable to corrosion by fresh mortar.
Best protection for lead components is a bitumen-based product (e.g. paint, tape).

Uses
D.P.C.s, radiation shields, paint pigments, sheathing of electric cables, ornamental work, solders, flashings, (occasionally) roof coverings.

Aluminium (Al)

Occurrence
Most common metallic constituent of the earth's crust, mainly found as bauxite ($Al_2O_3.2H_2O$).

Properties
Light (S.G. = 2.5–2.8).
Relative high cost of aluminium is due to the electrolytic method of extraction, but this permits very high purity.
Strength increases in aluminium alloys containing Mg, Mn, Si, Zn.
Good electrical and thermal conductivities.
Polished surface reflects heat and light.
High coefficient of thermal expansivity.
Non-magnetic.
Resistant to atmospheric corrosion due to formation of a protective oxide film. *Anodising* artificially increases this coating.
Liable to attack by alkalis.

Uses
Cladding, window frames, cables, vapour barriers, decorative metalwork, dip treatments.

Zinc (Zn)

Occurrence
As zinc blende (zinc sulphide).

Properties
Low tensile strength, but can be increased by alloying or hot working.
Low melting point.
Forms a good, corrosion-resistant coating for iron and steel. Alloying between base metal and zinc may occur.
Corroded by acids in timbers, e.g. oak.

Uses
Protection against corrosion, paint pigments, flashings.

CORROSION

Corrosion is the conversion of a metal to a non-metallic state.

Effects of corrosion on a metallic component

1. Structural soundness of component may be affected.
2. Growth of corrosion products on metal may cause cracking of material in which it is embedded.
3. May cause total failure of component.
4. May destroy aesthetic appearance of surface.

Types of corrosion

Dry oxidation
Surface metal atoms react with the oxygen molecules to form an oxide coating.
The rate of reaction depends on the type of metal and on temperature.

Electrochemical (wet) corrosion
Metals tend to ionise in aqueous solutions, releasing electrons; if this reaction can be maintained, total dissolution of the metal will occur.
The electrode potentials of some pure metals in solutions of their own ions, under standard conditions measured against a hydrogen electrode, indicate preferential rates of corrosion:

Metal	*Standard Electrode Potential* (volts)
Magnesium	−2.34 (anodic end)
Aluminium	−1.67
Zinc	−0.76
Iron (ferrous)	−0.44
Hydrogen (reference)	0.00
Copper (cupric)	+0.35
Gold	+1.68 (cathodic end)

48 METALS (3)

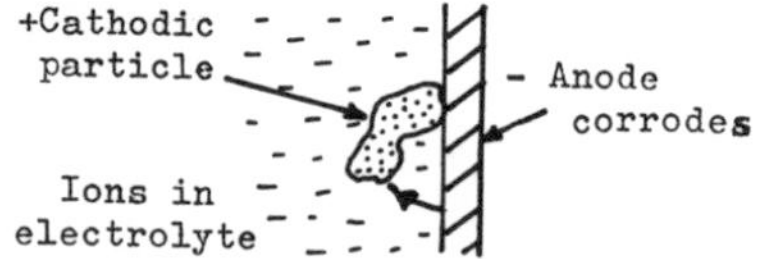

Corrosion of *two* dissimilar metals joined together in an electrolyte occurs mainly in the most anodic element.
Electrolytic corrosion in *one* metal can be caused by differential aeration.
'Stray' electric currents may also induce corrosion.

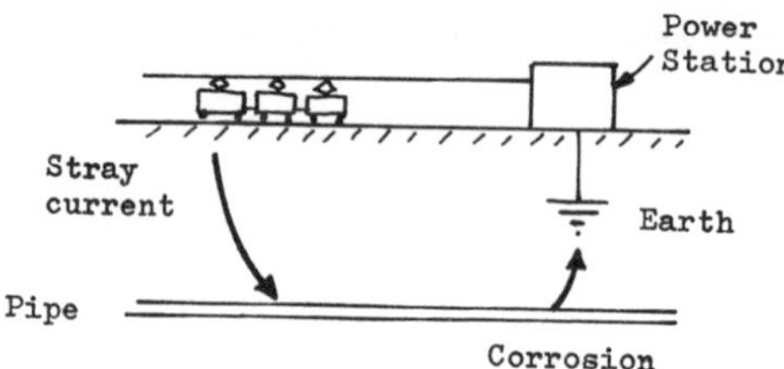

Factors influencing rates of corrosion

1. Composition of metal/alloy.
2. Temperature.
3. Design of plant or system.
4. Presence of organic matter and bacteria.
5. Presence of aqueous solutions, and their chemical properties.

Prevention of electrolytic corrosion

1. Sacrificial anode.
2. Impressed current.
3. Metallic coatings (e.g. galvanising, electroplating).
4. Paint coatings.
5. Thermoplastic coatings.
6. Vitreous enamel.
7. Anodising of aluminium and its alloys.
8. Inhibitors.

ALLOYS

Alloy: a multicomponent liquid or solid in which the primary component is a metal.

Cooling curves

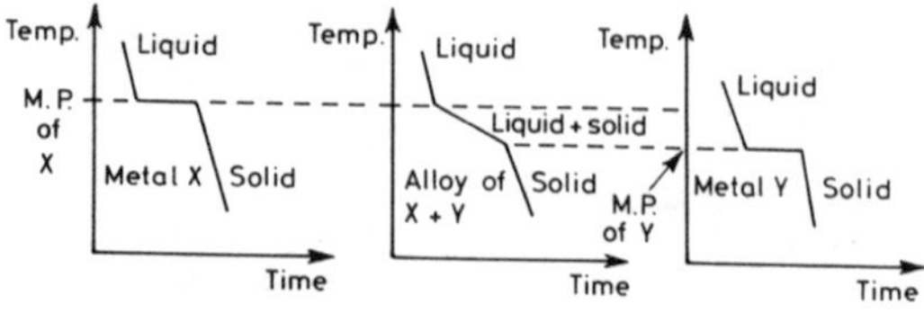

Equilibrium diagrams show which *phases* in a material are present, their composition and structure.

Gibbs phase rule

$$\boxed{P = C - F + 2}$$

where P is the number of phases which can coexist in equilibrium in a given system (solid, liquid, gas)
C is the number of components
F is the number of degrees of freedom (temperature, pressure, composition).

Two component system: solid solubility

Atoms of both components must satisfy the Hume-Rothery rules for complete solid solubility. Applying the Gibbs phase rule, $P = 2$ (constant pressure).

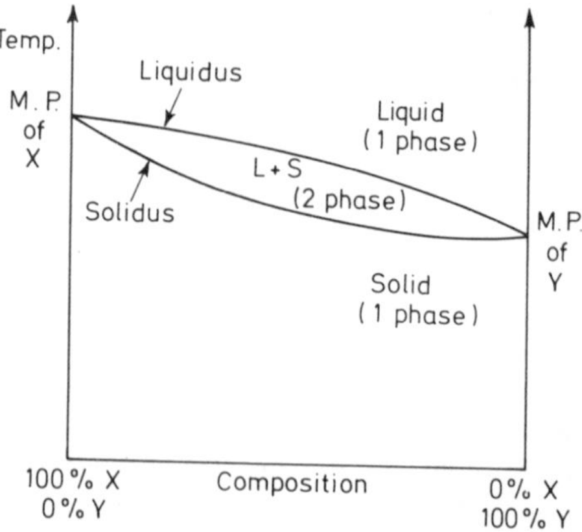

Lever rule is used to determine the proportion of each phase present at a given temperature and composition.

Two component system: limited solubility

Eutectic: Liquid $\xrightarrow{\text{cooling}} \alpha + \beta$
where α and β are solid phases.

Peritectic: Liquid $+ \alpha \xrightarrow{\text{cooling}} \beta$

PLASTICS

Plastics are organic materials which at some time in their history are capable of flow.

Formulation of compounds

Addition of *monomers* by the process of *polymerisation* to produce *polymers*. *Co-polymers* are chains of different monomers.
Addition polymerisation: 'butt-joining' of monomers.
Condensation polymerisation: 'dovetailing' of different monomers during a chemical change.

Types of plastics

Thermoplastics

1. Can be softened and resoftened indefinitely by the application of heat and pressure.
2. Structure consists of chains of monomers (M):
   —M—M—M—M—
   —M—M—M—M—
   Consist of *binder* (basic material), *filler* (for strength), plasticiser, colouring agent.
3. Vary from rigid to soft and pliable.
4. *Creep* under prolonged stress.
5. Vary from soft and easily scratched to hard and scratch resistant.
6. Suitable for production in continuous extruded form.
7. Include vinyl plastics, polyesters, polyamides, acrylics, fluorine plastics.

Thermosetting plastics

1. Undergo a chemical change when subjected to heat and pressure, and cannot be changed by further heating or pressure.
2. Structure consists of cross-linked chains of monomers (M):

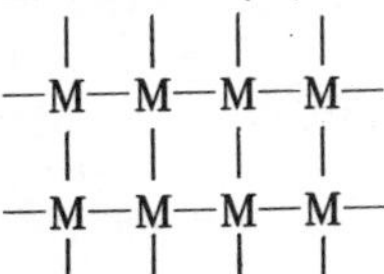

   Consist of binder, filler, plasticiser, colouring agent, *hardening agent* (to produce cross-linking), *accelerator* (to speed up action of hardening agent).
3. Rigid materials, except when foamed or in thin sheets.
4. Little affected by long-term stress.
5. Generally very hard and scratch resistant.
6. Not commonly available in continuous extruded form.
7. Include epoxy resins, polyester resins, amino resins and moulding powders, phenolic resins and powders.

Forming methods

1. Moulding (compression, injection and extrusion).
2. Pressing (simple and vacuum assisted).
3. Forming (vacuum, air pressure).
4. Calendering.
5. Casting.
6. Paste spreading and spraying.
7. Other methods including laminated and foamed plastics.
8. Fibre-reinforced plastics.

Joining methods

Screws, bolts, snap action, adhesives, heat or ultrasonic welding, solvent cementing.

Properties

Specific gravity 0.9 to 2.2

Strength
Low modulus of elasticity.
Favourable strength-to-weight ratios.

Thermal properties
High coefficients of thermal expansivity.
Low thermal conductivity of expanded plastics.
Degradation increases and strength decreases at high temperatures.

Electrical properties
Low electrical conductivity.
Attract electrostatic charges.

Combustibility
All plastics are combustible but some incorporate fire-retardant additives.

Acoustic properties
Acoustic insulation is generally poor.

Moisture resistance
Little water absorbed (except some nylons which swell).

Durability
Do not corrode or rot, although chemical breakdown increased by sunlight.

Economics
Favourable cost comparison with other building materials. Minimum worker education necessary.

Applications

Include pipes, rainwater goods, electrical insulation, cladding panels, roof lights, weather protection films, etc.
Major limitations to use for structural purposes, although may be incorporated in composite units, for inherently stiff structural geometries having low distributed stresses. *Glass reinforced plastics* (GRP) provide strong but not rigid structural skins.

BITUMINOUS PRODUCTS

Bitumens occur naturally as *asphalt*, or may be produced synthetically by *distillation of oil.*

Mastic asphalts

1. Consist of bitumen with inert mineral material —usually *graded, crushed limestone.*
2. *Fillers* added to reduce thermal movement.
3. *Uses*: damp-proof courses, paints, roof and floor coverings.

Coal tar

1. Runs at a lower temperature than bitumen.
2. *Uses*: tarmacadam for foot-paths.

Pitch

1. Residue left after distillation of tar from coal before all volatile fractions are removed.
2. *Uses*: roofing felt (incorporates mineral fibres).

Properties

Specific gravity 2.1

Strength
'Flow' under mechanical stress.

Thermal properties
Softened by heat and sunlight and over long periods will deteriorate. Exposed surfaces finished with light-reflecting material.

Combustibility
Pure materials are combustible but composite products are not easily ignited.

Moisture resistance
Excellent waterproofing qualities.

Durability
Very durable if not overheated.

Toxicity
Fumes from hot pitch are toxic.
Water in contact with some bituminous products can poison fauna and flora.

PAINTS

Functions

1. To protect surfaces from weathering, chemical, insect and fungi attack.
2. To modify surface properties.
3. To provide a decorative surface finish.

Constituents of paint

1. **Pigment**
   Fine, insoluble, organic or inorganic particles: e.g. titanium oxide (white), lamp black, zinc alloys, lead oxide, chrome yellow. If the resulting paint film is to be glossy, the pigment must be totally covered. Tests on pigments include light fastness, staining, obliterating power and oil absorption.
2. **Vehicle** comprises the following:
   Medium and binder: an oil (usually linseed) that forms the basis of paints.
   Resins: combine with the oil to produce a varnish.
   Extenders: substances (e.g. china clay) that improve paint viscosity, brushing qualities, provide a key for successive coats and reduce cost by acting as a filler. Extenders do not affect colour as their refractive index is similar to that of the medium.
   Solvents (thinners): volatile liquids that reduce the viscosity of the paint.
   Driers: accelerate the drying of oil-based paints, e.g. lead.
   Binders: improve the bonds between pigments and extenders during the hardening process.

Types of paint

1. **Oil paints**: the binder is an oil complex dissolved in a volatile solvent which evaporates when drying commences. Oxidation completes the drying process.
2. **Water paints**: supplied in paste form to which water is added. Drying occurs due to evaporation of water. Two types: washable and non-washable distempers.
3. **Emulsion paints**: a dispersion of pigments in a synthetic resin, e.g. polyvinyl acetate, emulsified in water. They do not contain organic solvents or oxidising agents and are not suitable for metallic surfaces.
4. **Thixotropic paints** ('non-drip'): gels that become more fluid when agitated or brushed on to a surface.

5. **Fire-retardant paints**: decompose under heat into substances that stifle combustion. *Intumescent* types produce carbon dioxide and water, and form an insulating cellular coating on the surface during combustion. *Halogenated* types produce toxic gases that reduce the spread of flame.
6. **Bituminous and rubber-based paints**: have anti-corrosive and water-repellent properties. Dry by the evaporation of a volatile solvent.
7. **Plastic/texture 'paints'**: consist of powdered gypsum, calcium carbonate and a binder. Usually applied with brush to ceilings and then textured. Finished decorative finishes tend to harbour dirt.
8. **Cement paints.** Oil paints are liable to attack by alkalis present in new Portland cement. Cement paints consist of cement, a waterproofer, an accelerator, and tinting pigments; they provide a durable waterproof finish for most porous surfaces. Flaking occurs if applied to brickwork containing soluble sulphates.

Application of paint coats

1. **Surface preparation**
   Ferrous metals should be descaled and cleaned by wire brushing, grit blasting, oxyacetylene flame or chemical treatment.
   Timber knots should be sealed to prevent resins leaking into the paint film.
   Bituminous products should be sealed to prevent bleeding and discoloration of paint coats.
   Holes and cracks should be filled with a water-resistant material—*stopper.*
2. **Primers**
   Provide adhesion for successive layers of paint.
   Primers for ferrous metals inhibit corrosion and should include red lead, calcium plumbate or zinc products.
   Primers for timber prevent rotting and should include red/white lead or be of the acrylic emulsion type.
3. **Undercoats**
   Provide opacity and correct colour density.
   One or more coats may be applied.
4. **Finishing coat**
   Acts as a weather seal.
   Provides the required finish in colour, properties and texture (matt, eggshell, semi- or full-gloss).
   Has a higher vehicle/solids ratio than undercoats but is slower drying.
   Gloss paints should never be applied to expanded polystyrene ceiling tiles because it increases the fire hazard.
   'Brilliant white' paints contain titanium chloride.

Paint failures

Bleaching: loss of colour due to acid/alkali attack.
Bleeding: breakdown of paint due to a chemical, e.g. resin exuded from an unsealed knot in timber, which reacts with the paint binder.
Blistering: due to trapped moisture or insufficient surface preparation.
Bloom: dimming of gloss sheen due to moisture in the atmosphere.
Chalking: a powder formed on the paint surface, caused by breakdown of the binder.
Cissing/pinholing: due to dirt particles.
Cracks: mainly found on external surfaces subjected to long periods of solar radiation.
Flaking: the breaking away of a complete coating from the surface due to incomplete adhesion.
Flotation: streaks of colour due to unsuitable combination of pigments.
Sagging: caused by excessive application of paint.
Saponification: results from the alkaline action between the damp substrate and the dried oil binder in the paint film, to form soap. It can be avoided by waiting for the plaster to dry out thoroughly or by using an alkali-resistant primer.

Removal of paint

By burning off (especially for painted timber).
By brushing on solvent paint-remover and scraping off paint film.
By abrasion or sand blasting (wire brushing not commercially used).

TIMBERS

Classes of commercial timbers

Hardwoods (Angiosperms)

Broad-leaved trees.
Deciduous or evergreen.
Seeds in pods or cases.
Examples: oak, elm, maple, mahogany.

Softwood (Gymnosperms)

Evergreen needles or scale-like leaves.
Coniferous trees, which are mostly evergreen.
Found mainly in northern temperate zone.
Seeds uncased.
Examples: redwoods (pines), whitewoods (spruces).
Note. Hardwoods are not necessarily more resilient than softwoods.

Nomenclature (refer BS 559 and 881)

1. Latin name, e.g. Pinus sylvestris.
2. Standard name, e.g. Redwood.
3. Common name, e.g. Red 'deal'.
   Note. Latin names preceded by an '×' indicate a hybrid plant, e.g. Quercus × kewensis.

Structure of timber

Trees have 3 main parts: *roots, trunk, crown.*

Roots
Anchor tree to the ground.
Take in water and mineral salts from the ground

Leaves
Act as a centre for *photosynthesis*:

$$CO_2 + H_2O + \text{mineral salts} \xrightarrow[\text{solar energy}]{\text{chlorophyll}} \text{food}$$

Trunk
Conducts water from roots to crown via fibro-vascular tissue.
Acts as a food store.
Provides support for leaves high above ground.

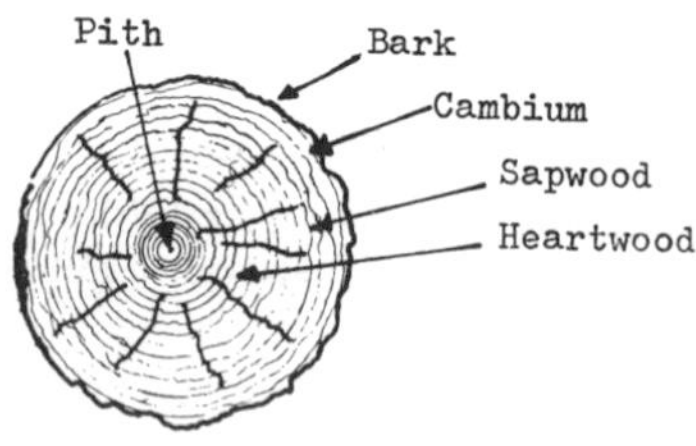

Outer bark
Provides protection against mechanical damage and extremes of temperature.

Inner bark
Part of translocation system.

Cambium layer
One-cell-thick continuous sheath that generates new cells.
Growth occurs in spring and summer in temperate zones, and a new layer of wood is added to the trunk and branches in one growing season. Cells big, with thin cell walls, at the beginning of the season, but smaller and darker growth formed later, resulting in *annual rings.*

Sapwood
Conducts water from roots to leaves.

Heartwood
Sapwood that has become compressed, stronger and usually darker in colour due to *tannins* and *gums* in *hardwoods* and *resins* and *coniferin* in *softwoods.*

Pith
Thin-walled cells.

Properties of timbers

Density
150 to 1250 kg/m^3 (depending on moisture content).
Softwoods have low densities.

Strength
In hardwoods *fibres* provide strength and *vessels* supply sap; in softwoods *tracheids* perform both functions.
Timber has a high strength/weight ratio.
Tensile strength *along* grain up to 30 times that *across* grain.
Tensile strength 2 to 3 times the compressive strength.
Strength increases with density of species.

Moisture content
Timber is *hygroscopic* when dry.
Moisture movement may cause shrinkage and swelling. Least shrinkage occurs radially.

$$\text{Moisture content (\%)} = \frac{\text{Wet weight} - \text{Dry weight}}{\text{Dry weight}} \times 100$$

Some 'green' softwoods have a moisture content $> 200\%$.
NHBC specify 9% to 17% for joinery, 17% for specialist timber work, 22% for carcassing.
Strength increases with decrease in moisture content.

Seasoning

Purpose is to reduce *moisture content* and *weight* of timber.

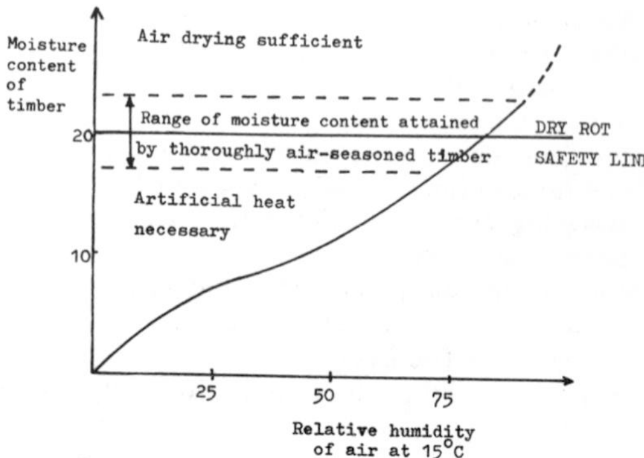

Methods of seasoning

Natural dry seasoning
Horizontal timbers separated by piling sticks stacked to allow natural air circulation over all surfaces.
Stack protected from direct sun and rain.
Rapid rate of drying will cause *case hardening.*
A bituminous compound may be applied to the ends of boards to prevent uneven drying.
Moisture contents of 15–19% attained.
Kiln seasoning
Artificial method to reduce seasoning time to days instead of months, without significant *degrade.*

Defects caused by seasoning

Cupping, bowing, twist, end splits, case hardening.

Stress grading

All structural timber must be stress graded (Building Regulations).

Principal stress grades

Visual inspection: Special Structural (SS) and General Structural (GS).
Machine grading equivalents: MGS and MSS; a higher grade M75 is also available.
Each piece of timber is inspected for:
knots, checks, shakes, wane, splits, rot, density* and slope of grain* (*cannot be detected by visual grading).

Deterioration of timbers

1. Fungal growth.
2. Insect attack.
3. Weathering.
4. Chemical attack.
5. Abrasion and mechanical damage.
6. Fire.

Fungal growth

Life cycle of a typical fungus. Sporophores (fruit bodies) discharge spores, which are carried on the wind to fall on damp wood. When the spore germinates *hyphae* are produced; these contain enzymes that are able to convert the cellulose of the wood into sugars, which nourish the fungus.

Favourable environments for growth
Damp wood (moisture content > 20%), suitable temperature, adequate oxygen.

Types of fungus
See table on page 54.

Insect attack

Life-cycle of a typical wood-boring insect. The adult beetle lays eggs in wood, which develop into *larvae*. The larvae, sometimes erroneously called *worms*, bore through the wood. The larvae remain dormant as *pupae* to emerge, usually in the summer, as adult insects.

Types of insect
See table on page 55.

Preservative

Classes of preservative

1. Class WB: Water-Borne type
2. Class TO: Tar Oil type
3. Class OS: Organic Solvent type

Application of preservatives
Brushing; spraying; immersion; 'full-cell' and 'empty-cell' processes. Diffusion process is used only for unseasoned wood.

Timber products

Plywood (BS 565 and 3493): an assembled product made up of plies and adhesives. It is stronger and stiffer than solid timber because the grains of adjacent plies are at right angles. Plies may be of different thicknesses.
Blockboard (BS 3444 and 3583): solid-core blocks faced with veneer whose grain is at right angles to the core grain.

Three ply

Blockboard

Chipboard (BS 1811 and 2604): wood particles bonded with synthetic resin.
Hardboard (BS 1142): partially dried wood pulp rolled to the required thickness.

Key to common wood-destroying fungi

The True Dry Rot fungus can attack wood with a lower moisture content than is generally required by the remaining fungi listed (wet rots). All, however, can attack wood with a wide range of moisture contents.

Name	*Where found*	*Wood attacked*	*External appearance*	*Fruit body*	*Appearance of wood after attack*
True Dry Rot *Merulius lacrymans*	Invariably in houses or buildings.	Softwoods, occasionally hardwoods.	(1) Fluffy white masses, often with yellow patches. (2) Matted grey skin, often tinged yellow or lilac. (3) Thin branching grey strands.	Flat pancakes. White edges, brick-red centre; frequently encountered in extensive outbreaks.	Dry and brittle, generally light brown. Deep cracks across the grain into brick-shaped pieces.
Cellar Fungus *Coniophora cerebella*	Very damp buildings, fences, sheds, etc.	Softwoods and hardwoods.	Frequently no external growth. Sometimes blackish brown strands on surface.	Rare in buildings. Thin flat plates, olive green or brown, usually with a pimply surface.	Dark brown, cracked along the grain. Often veneer of sounder wood on surface with cross cracking below.
Mine Fungus *Poria vaillantii*	Buildings, where very damp, also coalmines.	Softwoods.	Pure white branching strands like frosted window pane.	Flat white plates with visible pores.	Light brown, cracks across grain, similar to *Merulius* but less pronounced.
Paxillus panuoides	Sometimes in buildings.	Softwoods.	Dull yellow strands. Hairy and woolly.	Funnel or fan-shaped soft 'toadstool' growth. Yellowish.	Dark reddish brown, cracks with the grain, fine slits across grain.
Polystictus versicolor	Fence posts and window sills.	Hardwoods: sapwood and heartwood of non-durable species.	Often no external growth; occasionally whitish felted sheets.	Brackets, grey and brown zones on top.	Wood becomes lighter in colour.
Phellinus megaloporus (syn. *Phellinus cryptarum*)	Almost invariably in buildings.*	Oak.*	Little surface mycelium formed.	Hard and bracket-shaped; buff brown in colour.	White and stringy like lint in final stages; easily crushed but does not crumble to powder.
Poria xantha	Especially in greenhouses.	Softwoods.	Thin skin of yellowish white mycelium with small ill-defined strands. Not always present.	Thin sulphur yellow effused pore layer, may develop nodular outgrowths on vertical surfaces, which eventually become hard and chalky.	Brown with cubical cracking.

*Details refer to occurrence in the United Kingdom; in continental Europe also found on Chestnut and Oak in mines, caves, bridges, etc.

Key to common wood-boring insects

The details given below generally relate to average practical conditions; artificial or abnormal conditions can materially affect the life-cycles of insects.

Name	*Kinds of wood attacked*	*Number of eggs laid*	*Eggs hatch in*	*Length of pupal stage*	*Complete life-cycle*	*Adults emerge*	*Length of beetle*	*Size and shape of exit hole*	*Boredust*
Furniture beetle *Anobium punctatum* (Woodworm)	Hardwoods and softwoods, including ply-wood (containing natural glues).	Usually 20–40, but up to 80	4–5 weeks	4–8 weeks	2 or more years	April–September	2–5 mm	1.5 mm round	Roughly egg-shaped.
Death Watch Beetle* *Xestobium rufovillosum*	Hardwoods, occasionally softwoods close to hardwoods.	40–70 or more	2–8 weeks	2–4 weeks	Usually 4–5 years, can be up to 10	March–June	6–9 mm	3 mm round	Bun-shaped pellets.
Lyctus (Powder Post) Beetles Various species	Sapwood of certain hardwoods.	70–220	2–3 weeks	Up to 1 month	1–2 years, may be less in hot buildings	May–September	5 mm	1.5 mm round	Fine floury dust.
House Longhorn† *Hylotrupes bajulus*	Sapwood of softwoods, usually in roofing timbers.	Up to 200	2 or more weeks	About 3 weeks	3–11 years or more	July–September	10–20 mm	6–10 mm oblique slit	Short, compact cylinders and powdery dust.
Pentarthrum huttoni *Euophryum confine* (Weevils)	Hardwoods and softwoods, usually damp and decayed.	About 25 laid singly‡	About 2 weeks‡	6–8 weeks‡	7–9 months‡	No fixed time	3–5 mm	Channels on surface and some holes 1.5 mm	Similar to Anobium but smaller.
Termites	Do not occur in Great Britain								

*Emits a tapping sound as a mating call.

†Found mainly in S.E. England. Refer Building Regulations 1976, B4: softwood timbers used in a roof, including ceiling joists, in the local authority areas designated by the Building Regulations must be adequately treated with a suitable preservative to prevent this kind of infestation.

‡These data apply to Pentarthrum at 25°C and 95–100% RH.

ADHESIVES

Types

Animal glues

Derived from bones and hides.
On application, gelation and hardening occur.
Moisture tends to soften the glue.
Suitable for internal use, e.g. wood joints.

Casein glues

Obtained from milk.
Settling occurs by natural gelation and evaporation of water.
Suitable for fixing plasterboard to wood.

Thermoplastic adhesives

Setting occurs by cooling, solvent evaporation, or emulsion coalescence. Can be resoftened by heat.

Polyvinyl acetate (PVA) glues
Emulsify in water to become water resistant.
Do not require addition of a hardener.
One surface to be joined should be absorbent.
Used for timber joints and as a bonding agent between old and new concrete.

Thermosetting adhesives

Suitable for external use, owing to high moisture resistance.

1. **Amino resins**
   (a) *Urea formaldehyde*
   Twin-pack glues, one of which contains a hardener.
   Pressure must be applied to new joints.
   Crazing of the *glue line* will occur if a large gap occurs between surfaces to be joined.
   (b) *Melamine formaldehyde*
   Available in powdered form, to which water must be added.
   Very durable.
2. **Phenolic resins**
   Phenol formaldehyde
   Used for joining timber products to concrete and asbestos sheets.
   A gap-filling grade should be used on uneven surfaces.
3. **Epoxy resins**
   Produced in twin-pack and liquid form and provide good adhesion between most materials.
   Highly flammable and waterproof.
   Used as a contact adhesive.

Rubber-based adhesives

Good impact properties.

Bitumen-based adhesives

Good chemical and water resistance. Available in solvent or emulsion form.
Used in laying wood-block flooring.

MASTICS

Mastics are used to seal joints, fill gaps, provide bedding for window frames and external door frames, curtain walling.

Performance characteristics (only some may be necessary)

Adhere to surfaces; resist chemical attack; have suitable strength properties; should not crack, nor exude or slump on weathering; aesthetic requirements.

Types

Non-setting mastics

Exhibit plastic flow; age and weather-harden with time.

1. **Bituminous mastics**
   Very viscous.
   Applied hot, but slight over-heating can alter its properties.
   Cannot be successfully painted over.
2. **Oil mastics**
   Consist mainly or vegetable oils and filler.
   Provide a waterproof seal.
3. **Polyisobutylene mastics**
4. **Butyl mastics**

Setting mastics

1. **Polysulphide rubbers**
   Usually supplied in twin-pack form (base and curing agent).
2. **Silicone sealants**

Note. Building mastics may be used as adhesives to fix tiles and glass.

GASKETS

Preformed gaskets of rubber or plastic form dry-seal joints that rely on compressive forces, not adhesion.

RADIOLOGICAL METHODS

X-ray radiography

Short-wave electromagnetic radiation produced by an X-ray tube is passed through a specimen and recorded on photographic film. Fluoroscopic methods which convert X-rays to visible light are also available, but do not provide such good sensitivity as film.
Principle: depends on the differential absorption of radiation passing through the specimen, caused by density variations or different path lengths.
Application: testing welds and castings.

γ-ray radiography

Electromagnetic radiation with shorter wavelengths than X-rays (higher energy). Produced by decay from radio-active sources, e.g. Cobalt 60, Caesium 137, Iridium 192. Require longer exposure times than for X-rays.
Application: testing welds and castings; back scattering and transmission gauges; measurement of coating thickness and composition; radio-active sources can be used for leak detection.

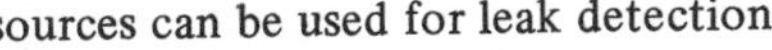

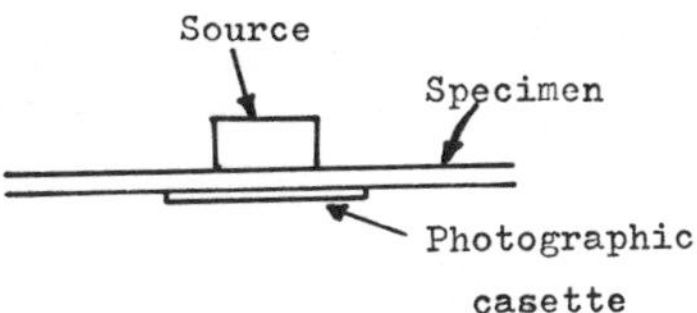

Advantages: smaller set up; low capital outlay; portable compared to most X-ray machines (although small, transportable, lightweight X-ray sets are available).

Limitations of radiographic methods

Size of defect detected; defect parallel to beam is not easily observed; provision of protection for personnel; γ-sources are not suitable for testing non-metallic specimens or light alloys; radiography is ineffective in detecting sharp cracks.

ELASTIC METHODS

Ultrasonic methods

Below 20 Hz: infrasonics
20 Hz to 20 kHz: audible range
Above 20 kHz: ultrasonics
Longitudinal (compression) waves are generated by a piezoelectric transducer, i.e. crystals change size when a voltage is applied across them.
Principle: the transducer is passed over the surface of the specimen and echoes are detected as electri-

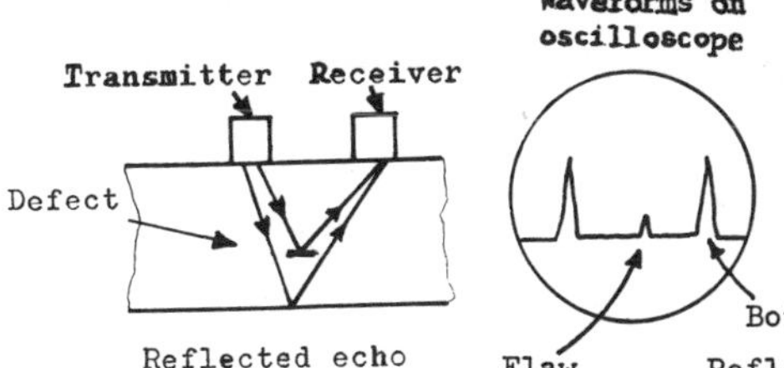

cal signals by a receiver, amplified and displayed on an oscilloscope. In solids, a second mode of vibration may be generated–transverse (or shear) waves. A third mode of wave motion, a Rayleigh surface wave, may also be used in testing for surface defects. Wavelength range 1–6 mm.
Flaw detection depends on the impedance of the defect, e.g. an air-filled crack in steel is easily detected but a slag inclusion will be partially transparent to ultrasound. Matched impedances between the probe and the specimen can be obtained with grease or water.
Application: inspecting welds in steel pipes, bridges etc.; finding crushing strength of concrete.
Advantages: portable; minimal health hazard; immediate results.
Limitations: null point near the transducer; sudden changes in shape cause spurious results; a rough surface introduces coupling inaccuracies; planar defects lying in direction of beam will not be detected.

Electrodynamic methods

Principle: the specimen is excited by a variable-frequency (audio-range) vibration, and frequency at which resonance occurs noted. This resonant frequency depends on the physical properties and dimensions of the specimen, hence the clamping position is critical.
Application: determining dynamic Young's modulus or shear modulus.

MECHANICAL METHODS

Schmidt rebound hammer

A 'rebound number' is obtained from the rebound of a spring-loaded hammer. This number can be converted into compressive-strength values using conversion graphs.
Application: determining compressive strength of concrete.

Indentation methods

Tests relate hardness to the resistance to plastic deformation.
Application: Vickers hardness-testing machine.

MAGNETIC PARTICLE METHODS

Principle: defects cause distortion of magnetic flux lines in magnetised specimens. The induced magnetic field should not be too low or high.
Application: detecting surface and sub-surface (up to 2 mm) cracks in ferro-magnetic compounds, e.g. plate and weld runs.

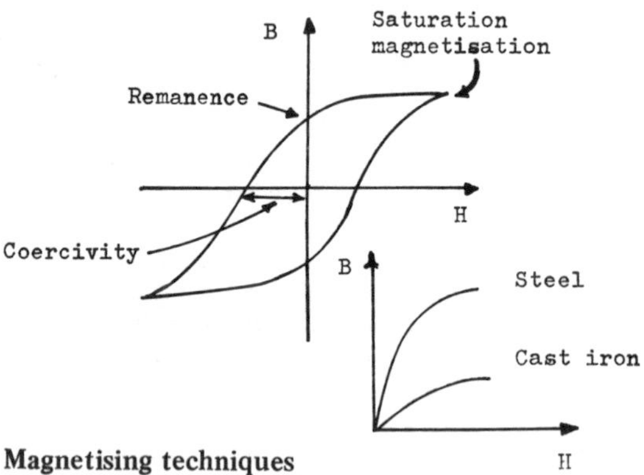

Magnetising techniques

Magnetic flux method: magnetic field is produced by permanent magnets or electromagnets and is used to detect defects perpendicular to lines of flux.

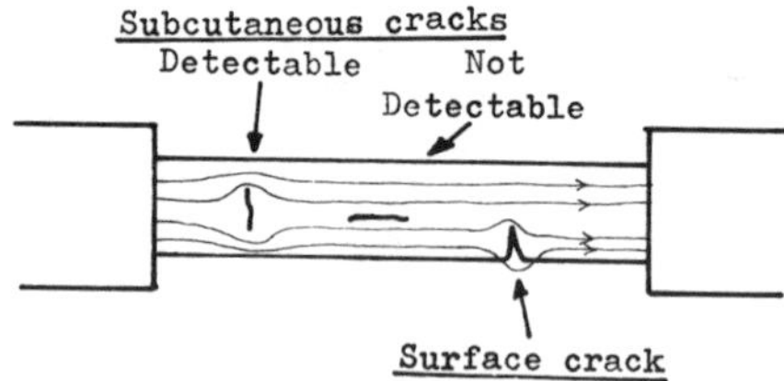

Current flow method: uses mains electricity to produce flux lines perpendicular to the direction of current flow and thus detects longitudinal cracks. Specimen forms part of electrical circuit.

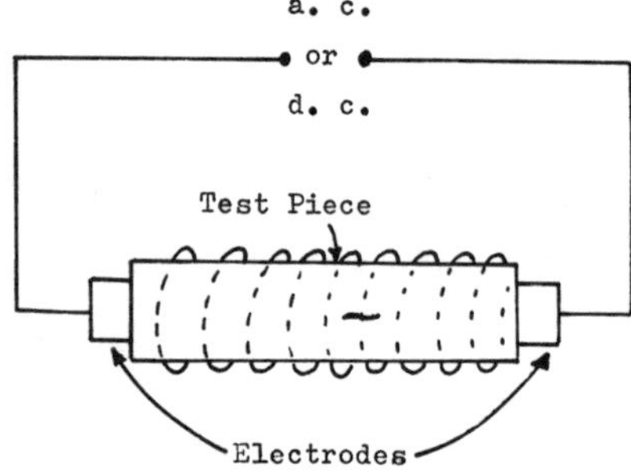

Threading bar method: uses mains electricity and detects longitudinal cracks. The specimen is *not* part of the electrical circuit.

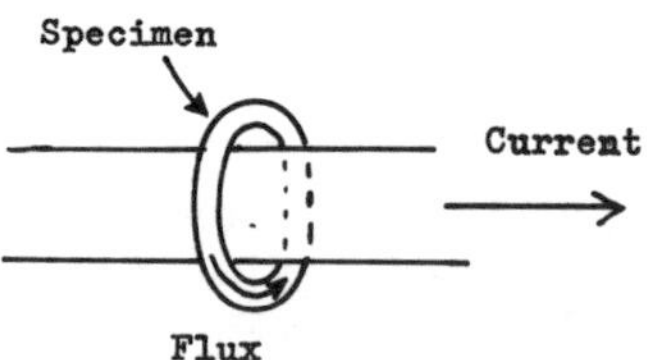

Note. Heat is generated in electrical methods. Magnetic ink or dry powder is used as an inspection medium. Black inks for polished surfaces and white for dark surfaces. Permanent records can be made by placing transparent self-adhesive tape over dry ink. The specimen should be demagnetised after the test.

ELECTRICAL METHODS

Resistance measurements, e.g. strain gauges

The electrical resistance of some metals changes according to their state of strain. Gauges, usually in pairs, are incorporated into the arms of a balanced Wheatstone bridge. Bonded gauges are fixed to the test object with a special cement which is temperature dependent.

D.C. search-coil methods

Leakage of flux from magnetic specimens can be detected by moving a search coil over the surface.

A.C. methods (up to radio frequency)

More sensitive than d.c. methods and depend on:
(a) out-of-balance signal produced in a 'bridge' circuit;
(b) eddy-current test methods.

THERMAL METHODS

Principle: all objects emit infra-red radiation, which can be detected using special photographic films in which final colours are temperature dependent.
Application: detecting heat losses from buildings.

PENETRANT METHODS

Principle: increases the contrast between the defect and its background.
Application: finding surface flaws in almost any solid material.
Types of penetrant
(a) Hot oil and chalk.
(b) Red dye: test kit consists of 3 aerosol components, i.e. cleaner, penetrant and developer.
(c) Fluorescent dye: viewing done under 'black light', i.e. near UV radiation.

Structure used to:

1. Control fire within a building (or without).
2. Provide safety for occupiers of building in case of fire.
3. Provide protection for property and contents.

Fire protective properties of materials:

1. **Non-combustibility test** (BS 476, Part 4)
   To observe whether a sample produces a flame in a special furnace (temperature continuously recorded).

   3 specimens, 40 mm × 40 mm × 50 mm, are prepared at 60°C for 24 hours.

   A material is deemed *combustible* if:
   (a) the temperature from either of the 2 thermocouples rises 50 deg C above the furnace temperature;
   (b) the specimen flames continuously for 10 seconds or more.
   Otherwise, the material is deemed *non-combustible.*

2. **Ignitability test** (BS 476, Part 5)
   To study the behaviour of materials when subjected to a small flame.

   3 specimens, 228 mm × 228 mm of normal thickness, are conditioned with air at 10–21°C and 55–65% relative humidity.

   If any specimen flames for more than 10 seconds after removal of test flame, specimen is *easily ignitable* (X).
   If no specimen flames for more than 10 seconds, specimen is *not easily ignitable* (P).

3. **Fire propagation test** (BS 476, Part 6)
   To determine a numerical index relating to the amount and rate of heat evolved by a specimen whilst being subjected to heat in an enclosed space under prescribed conditions.

4. **Surface spread of flame test** (BS 476, Part 7)
   Large-scale test to determine the tendency of materials to support flame across their surface.

   6 specimens, 230 mm × 900 mm of normal thickness (not exceeding 50 mm) conditioned at 10–21°C and 55–65% relative humidity. (Small-scale test is also specified.)

   Class 1 (very low) to Class 4 (rapid flame spread).
   Class 0 (Building Regulations E15):
   either the material for ceilings or walls is construed to be non-combustible throughout,
   or the surface material conforms to the appropriate stated values when tested in accordance with BS 476, Part 6: 1968.

5. **Fire-resistance test** (BS 476, Part 8)
   To assess the ability of elements of construction to retain their stability, resist passage of flame, hot gases and heat transmission.

   Test is made on a representative of the element of construction.
   (a) *Stability*
   Non-load bearing: failure is deemed to occur with collapse of specimen.
   Load bearing: specimen should support test load during, and for 24 hours after, heating period.
   (b) *Integrity*: failure occurs when cracks etc. appear.
   (c) *Insulation*: failure occurs when the mean temperature of the unexposed surface of the specimen increases by more than 140°C above the initial temperature, or by 180°C at any point on that surface.

6. **External fire exposure roof tests** (BS 476, Part 3)
   Tests designed to show how a roof will behave when subjected to radiated heat and burning brands from a fire in a nearby building.

   The result of the tests on specimen roof coverings is expressed by a two-letter designation, e.g. AA, AB, BB. The first letter denotes the time the specimen resisted penetration by radiated heat from an outside source; the second letter denotes the extent of spread of flame along an external surface. Highest classification is AA, e.g. galvanised steel.

   The suffix X is added to the designation if dripping from underneath, mechanical failure, or holes occur in the test specimen; e.g. ACX.